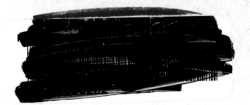

Developments in Solar System- and Space Science, 3

Editors: Z. KOPAL and A.G.W. CAMERON

Developments in Solar System- and Space Science, 3

LUNAR STRATIGRAPHY AND SEDIMENTOLOGY

by

JOHN F. LINDSAY

The Lunar Science Institute, Houston, Texas, U.S.A.
and
The Marine Science Institute, The University of Texas, Galveston, Texas, U.S.A.

ELSEVIER SCIENTIFIC PUBLISHING COMPANY

AMSTERDAM — OXFORD — NEW YORK   1976

ELSEVIER SCIENTIFIC PUBLISHING COMPANY
335 Jan van Galenstraat
P.O. Box 211, Amsterdam, The Netherlands

AMERICAN ELSEVIER PUBLISHING COMPANY, INC.
52 Vanderbilt Avenue
New York, New York 10017

The camera-ready copy for this book has been prepared by
Medical Illustrations Services of The University of
Texas Medical Branch, Calveston, Texas, U.S.A.

ISBN: 0-444-41443-6

Printed in The Netherlands

TO

The Apollo Astronauts.

The moon is nothing
But a circumambulatory aphrodisiac
Divinely subsidized to provoke the world
Into a rising birthrate.

The Lady's Not For Burning
Christopher Fry, 1950

Preface

The dominant processes operative in shaping the lunar surface are very different from those acting on the earth's surface. The main differences between the two planets relate to the way in which available energy is used in the sedimentary environment. Sedimentary processes on the earth's surface are determined largely by solar energy interacting with the atmosphere and hydrosphere which act as intermediaries converting radiative solar energy to effective erosional and transportation energy by way of rivers, glaciers, ocean waves and so on. The moon is essentially free of both an atmosphere and hydrosphere and as a consequence solar energy is largely ineffective in the sedimentary environment. Instead lunar sedimentary processes are dominated by kinetic energy released by impacting meteoroids.

In the early stages of the Apollo program considerable attention was given to locating landing sites which would provide the best opportunity of sampling the primitive lunar crust. As the Apollo program progressed it became apparent that most of the rocks available at the lunar surface were in fact "breccias" or "clastic rocks" or in a more general sense "sedimentary rocks." The moon's crust was much more complex than anyone might have guessed. This book is an attempt to organize some of the information now available about the sedimentary rocks forming the lunar crust in a way that allows some comparison with the terrestrial sedimentary environment.

There are essentially three parts to the book. Chapter 1 presents a very brief view of the moon as a planetary body to establish a perspective for the following chapters. Chapters 2 and 3 evaluate the energy sources available in the lunar sedimentary environment. Because of their predominance in the lunar environment meteoritic processes are treated in considerable detail. Chapters 4, 5 and 6 bring together information on the general geology of the lunar crust and detailed information from some sedimentary units sampled during the Apollo missions.

A large number of people have contributed in various ways to make it possible for me to write this book and I am grateful for their assistance. In the early stages Dr. J. Head, then acting director of the Lunar Science Institute, Houston, Texas, encouraged me to begin the book and Prof. Alan

White, La Trobe University, Melbourne, Australia, generously allowed me considerable time away from teaching duties. Much of the work on the book was done while I was a Visiting Scientist at the Lunar Science Institute and I am especially grateful to Dr. R. Pepin the director of that institute for his encouragement and for considerable support in the preparation of the manuscript and diagrams. Many other staff members of the Lunar Science Institute provided invaluable help particularly Ms. F. Waranius and Ms. G. Stokes who sought out numerous obscure references for me and gave me unlimited access to excellent library facilities and Ms. M. Hagar who helped find many lunar photographs and Ms. C. Watkins who coordinated much of the drafting and photography for me. Without the precise typing and inspired interpretation of the handwritten drafts by Ms. L. Mager of the Lunar Science Institute and Ms. C. Castille of the University of Texas at Galveston, the manuscript may never have been read by anyone but the author. I am grateful to Mr. B. Mounce, NASA, Johnson Space Center and Ms. C. Martin, University of Texas for the drafting of most of the diagrams and Mr. R. Henrichsen and Ms. R. List, University of Texas at Galveston who coordinated the preparation of camera-ready copy prior to publication. The National Space Science Data Center provided some Lunar Orbiter photography.

Finally, I would like to thank Dr. G. Latham, Associate Director of the Marine Science Institute, University of Texas at Galveston for his continued encouragement particularly during difficult times in the final stages of writing and Dr. D. McKay, Dr. F. Hörz and Capt. J. Young who provided helpful advice during preparation of the manuscript.

January 12, 1976   John F. Lindsay

Contents

A composite view of the lunar nearside. The mare and highlands are clearly differentiated and the major circular basins are visible. (1) Mare Orientale on the edge of the disk. (2) Oceanus Procellarum. (3) Mare Humorum. (4) Mare Nubium. (5) Mare Imbrium. (6) Mare Serenitatis. (7) Mare Tranquillitatis. (8) Mare Nectaris. (9) Mare Fecunditatis. (10) Mare Crisium. (Lick Observatory Photograph).

CHAPTER 1

The Moon as a Planet

Introduction

Even though a satellite of the earth, the moon is a planet and is not much smaller than most of the terrestrial planets (Table 1.1). The Apollo missions thus returned with the first rock samples from another planetary body and provided us with a second data point in our study of the solar system. These samples tremendously increased our knowledge and understanding of the evolution of planetary bodies. Possibly just as significant as the returned samples was the variety of instruments set up on the lunar surface to monitor the moon's surface environment in real time. Particularly important in this respect were the lunar seismometers, magnetometers and the heat flow experiments. Also of importance in the synoptic view of the moon are a series of orbital experiments, particularly the orbital measurement of X-rays excited by solar radiation and γ-rays produced by natural radioactivity on the lunar surface which provide data on the areal distribution of some elements.

The Apollo program has left us with a picture of the moon as essentially a dead planet with most of its internal energy having been expended in the first 1.5 aeons (1 aeon = 10^9 yr) of its history. However, during this early period the moon appears to have been very active. As its internal structure evolved its face was dramatically altered by giant meteoroid impacts that created an extensive and complex crustal stratigraphy. Finally, the basins excavated by these impacts were themselves flooded on a large scale by basaltic lavas to form the familiar lunar nearside mare.

Gilbert (1893) and Shaler (1903) were perhaps the first to seriously consider the geology of the lunar surface and to attempt to explain what they saw in terms of natural processes. However, it was not until Shoemaker and Hackmann published their studies in 1962 that the geology of the moon was established on a firm foundation. For the first time the moon was seen as being somewhat similar to the earth in that it had a stratigraphy in which at least the later history of the development of the planet was recorded.

As a result of the Apollo program, the structure of the moon is now

TABLE 1.1.

Physical parameters of the moon compared to the sun and planets.

Terrestrial Planets

	Moon	Sun	Mercury	Venus	Earth	Mars
Distance from sun (millions of km)	150	—	58	108	150	229
Mass (units of 10^{27} g)	.0735	1,984,000	0.3244	4.861	5.975	0.6387
Density (g cm^{-3})	3.34	1.41	5.45	5.15	5.52	3.97
Volume (units of 10^{27} cm^3)	0.02199	1,410,000	0.06	0.06	1.08	0.16
Radius (km)	1738	696,000	2,400	6,100	6,371	3,400
Gravity (cm s^{-2})	162	—	360	850	982	376

Outer Planets

	Jupiter	Saturn	Uranus	Neptune	Pluto
Distance from sun (millions of km)	778	1,430	2,860	4,490	5,910
Mass (units of 10^{27} g)	1,902	569.4	86.88	102.8	0.3 ?
Density (g cm^{-3})	1.33	0.68	1.60	2.25	4 ?
Volume (units of 10^{27} cm^3)	1,410	802	55.1	41.6	0.15
Radius (km)	69,900	57,000	26,000	22,000	2,900 ?
Gravity (cm s^{-2})	2,600	1,120	940	1,500	800

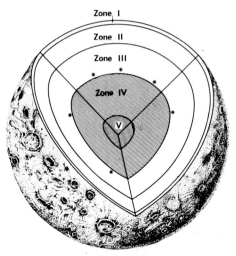

Fig. 1.1. A cutout view of the moon illustrating the various zones described in the text and showing the locations of moonquakes. (from Nakamura *et al.*, 1974).

known in considerable detail. The following sections briefly consider the overall structure of the planet and its possible origins. Later chapters look more closely at the stratigraphy of the lunar crust.

Internal Structure and Chemistry of the Moon

The accumulated geophysical and geochemical data resulting from the Apollo missions places constraints on the structure and composition of the moon and ultimately on its origin and evolution (Table 2.1). On the basis of the lunar seismic data Nakamura *et al.* (1974) proposed a lunar model consisting of five distinguishable zones (Figs. 1.1, 1.2 and 1.3).

The Crust (Zone I)

The moon possesses a differentiated crust which varies from 40 km thick at the poles to more than 150 km thick on the lunar farside (Wood, 1972). On the lunar nearside the crust is 50 to 60 km thick and is characterized by relatively low seismic velocities (Fig. 1.2). The outer kilometer of the crust has extremely low velocities, generally less than 1 km-1, that correspond to lunar soil and shattered rock. The velocities increase rapidly to a depth of 10 km due to pressure effects on the fractured dry rocks (Latham *et al.*, 1972). Seismic velocities to a depth of 25 km in Oceanus Procellarum coincide with measured velocities of mare basalts. Below 25 km the velocities increase slightly suggesting a transition to the

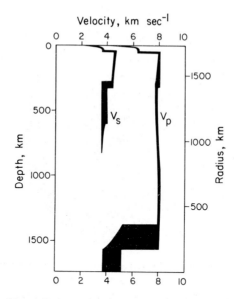

Fig. 1.2. Seismic velocity profiles of the lunar interior. (from Nakamura *et al.*, 1974).

anorthositic gabbro which is characteristic of the highland lithologies.

The lunar crust consists of an anorthositic gabbro ($\approx$ 70 per cent plagioclase) together with smaller amounts of a second component which is essentially a high alumina basalt enriched in incompatible elements and referred to as Fra Mauro basalt or KREEP (indicating an enrichment in potassium, rare earth elements and phosphorous). Locally, the crust is overlain by a variable thickness of mare basalt. The outer portions of the crust are extremely inhomogeneous due to differentiation at shallow depths and the fact that large segments have been overturned by the 40 or more ringed-basin forming events (Stuart-Alexander and Howard, 1970; Cadogan, 1974).

The source of the highland crustal rocks is considered to be an alumina-rich layer at shallow depths. Two hypotheses are available to account for this chemical zonation (Taylor *et al.*, 1973): ·

    1.   The outer 200-300 km was accreted from more refractory alumina-rich material than was the interior.

    2.   An alumina-rich outer crust was formed by melting and differentiation early in the history of the moon, following homogeneous accretion. The well developed positive K/Zr and K/La correlations observed both in mare basalts and highland samples supports this argument (Erlark *et al.*, 1972; Duncan *et al.*, 1973; Church *et al.*, 1973; Wanke *et al.*, 1973). In the outer layers of the moon were accreted from more

refractory material than the interior, it is difficult to account for this correlation between the volatile element, potassium, and the refractory element, zirconium; and the heterogeneous accumulation model faces a serious problem.

Experimental investigations (Green, 1968; Green *et al.*, 1972) show that the formation of anorthositic gabbroic magmas requires rather extreme conditions. It appears likely that the uppermost plagioclase-rich layer of the lunar highlands represents a cumulate and is underlain by a layer relatively depleted in plagioclase — possibly a pyroxene-rich gabbro. The average composition of the lunar crust would thus be gabbroic of composition which could form by advanced partial melting of source material, estimated to contain 4 per cent to 5 per cent of Al_2O_3 (Ringwood, 1970a). Assuming a content of 13 per cent Al_2O_3 for the parental cotectic gabbro, the existence of a crust some 65 km thick would require complete differentiation of a layer at least 200 km deep (Duba and Ringwood, 1973). The Fra Mauro basalt or KREEP component is interpreted as formed by a very small degree of partial melting at even greater depths (Ringwood and Green, 1973). Whilst the thickness of the crustal layer may vary, it is difficult to avoid the conclusion based on abundances of Al and U in the lunar highlands that the differentiation event which took place ≈ 4.6 aeons ago involved 30 to 50 per cent of the moon's volume.

The Upper Mantle (Zone II)

The upper mantle is approximately 250 km thick and is characterized by P wave velocities of about 8.1 km s^{-1} at the top and possibly slightly decreasing velocities at depth.

Experimental petrology indicates that the mantle is the probable source region for the mare basalts and that it was probably composed largely of orthopyroxene and sub-calcic clinopyroxene together with some olivine (Ringwood and Essene, 1970; Walker *et al.*, 1974). An olivine-pyroxene

TABLE 1.2.

Composition of the Moon (adapted from Taylor and Jakes, 1974).

	SiO_2	TiO_2	Al_2O_3	FeO	MgO	CaO
Bulk Moon	44.0	0.3	8.2	10.5	31.0	6.0
Crust	44.9	0.56	24.6	6.6	8.6	14.2
Upper Mantle	43.3	0.7	14.4	12.6	17.7	11.3
Middle Mantle	42.9	——	0.4	13.5	42.7	0.5

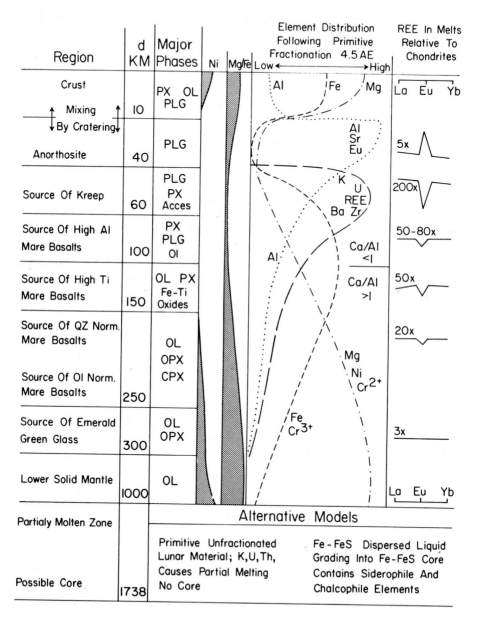

Fig. 1.3. Geochemical model of the lunar interior at about 4.4 aeons (Taylor and Jakes, 1974, *Proc. 5th Lunar Sci. Conf., Suppl. 5, Vol. 2*, Pergamon).

mineral assemblage is consistent with the seismic data although compositions richer in olivine near the top of the zone are more probable (Nakamura *et al.*, 1974). The total volume of the mare is estimated to be only 0.2 per cent of the moon (Baldwin, 1970). If the mare basalts represent liquids formed by 5 per cent partial melting of their source regions the total volume of differentiates would be 4 per cent of the lunar volume (Duba and Ringwood, 1973). If the moon was completely differentiated 3 aeons ago the mare basalts represent only a small proportion of the liquids formed. A large volume of magma thus failed to reach the surface and crystallized as plutonic intrusions in the outer cool lithosphere of the moon (Baldwin, 1970; Ringwood, 1970; Duba and Ringwood, 1973). This is predictable since the density of molten mare basalt is similar to that of the lunar crust. Consequently an excess is necessary for lava to be extruded at the surface. Most of the magma will be extruded as sills, laccoliths and batholiths in the crust or as large plutons at the crust-mantle boundary.

Upon cooling the lunar mare basalts pass into pyroxenite and eclogite stability fields at lower pressures than terrestrial basalts (Ringwood and Essene, 1970; Ringwood, 1970a; Green *et al.*, 1971). Consequently large plutons emplaced in the upper mantle were probably transformed into pyroxenite ($\rho \approx 3.5$ g cm^{-3}) and eclogite ($\rho \approx 3.7$ g cm^{-3}) resulting in a non-uniform density distribution in the outer parts of the moon and possibly explaining the existence of mascons.

The Middle Mantle (Zone III)

The middle mantle corresponds to a zone of reduced S wave velocities from 300 to 800 km in depth. Deep moonquakes occur at the base of this zone suggesting that the moon is relatively rigid to the base of the middle mantle. This zone may consist of primitive lunar material that lay beneath the zone of initial melting that resulted in the formation of the crust and upper mantle. Conductivity studies suggest that large volumes of the lunar interior at depths of several hundred kilometers are at temperatures close to the solidus for lunar pyroxenite (Fig. 1.4) (Duba and Ringwood, 1973).

The Lower Mantle (Zone IV)

The zone below 800 km in depth is characterized by high attenuation of shear waves. The lower mantle thus may be partially molten and similar to the asthenosphere of the earth (Nakamura *et al.*, 1974).

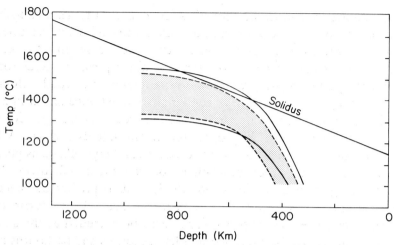

Fig. 1.4. Selenotherms based on conductivities (Duba and Ringwood, 1973; Sonett *et al.*, 1971). The solidus of a lunar interior model (Ringwood and Essene, 1970) is plotted for reference. Selenotherms calculated from data on pyroxene conductivity fall within the stippled area, those calculated from data on the conducitivity of olivine fall within the lined areas.

The Core (Zone V)

There is an inner zone of reduced P wave velocities between 170 and 360 km in diameter which, by analogy with the earth, could be an iron-rich core. Iron is the only abundant element that can be expected to occur as a metal and it satisfies the requirement for a dynamo within the moon (Brett, 1973). A lunar core might also be expected to contain nickel otherwise the moon must be depleted in nickel with respect to chondritic abundances. However, the thermal restraints imposed by a molten core of pure iron or nickel-iron mixture are not consistent with other lunar data (Brett, 1972, 1973). By extrapolating from experimental data it can be shown that the melting point of metallic iron at the center of the moon is approximately 1660°C (Brett, 1973; Sterrett *et al.*, 1965). The presence of nickel lowers the melting point only slightly. Temperatures of this order at the center of the moon early in its history are not consistent with current thermal models (Toksoz *et al.*, 1972; McConnell and Gast, 1972; Hanks and Anderson, 1969; Wood, 1972). The melting curve for a pure-iron core lies close to or below the solidus curves for most reasonable models of the lunar mantle (Fig. 1.5). A molten iron core would require that a large proportion of the moon be at or close to its melting point at least 4.1 to 3.1 aeons ago.

One alternative to a nickel-iron core is a molten Fe-Ni-S core (Brett, 1972, 1973). Sulfur alloys readily with iron and it has been found that sulfur and other volatile elements are strongly depleted in lunar basalts. The Fe-FeS eutectic temperature at 1 atmosphere is 988°C (Hansen and Anderko, 1958). The eutectic temperature appears to be changed little by increased pressure, but the eutectic composition becomes more Fe-rich. The eutectic temperature at 50 kbars (the approximate pressure at the center of the moon) is approximately 1000°C at a composition of $Fe_{75}S_{25}$ (weight per cent) (Brett, 1973). The pressure of sulfur in the core thus allows a molten core to exist at temperatures close to 1000°C (Fig. 1.5). This would be further lowered by the presence of nickel. A core containing pure sulfur would also contain other chalcophile elements such as Pb, Cu, Bi, Se, Te and As, many of which are also depleted in the lunar basalts. A core containing sulfur has a lower density than a pure iron core and is much more consistent with the constraints imposed by the moons moments of inertia. A core occupying 20 per cent of the moon's radius would require a bulk sulfur content of only 0.3 weight per cent. A core of these dimensions could be derived during the major differentiation of the moon from one half of the moon's mass (Duba and Ringwood, 1973). The molten metallic phase may have segregated during the major differentiation of the moon that occured 4.6 aeons ago (Duba and Ringwood, 1973). It probably collected in a layer at a depth where the silicate mantle was viscous enough to impede the fall of this dense metal phase toward the center. This layer was gravitationally unstable and any irregularity in the metal-silicate interface may have served as a nucleation point for the formation of a drop of metal. This drop grew expotentially and eventually fell to the center.

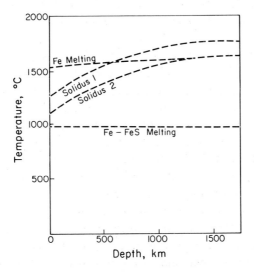

Fig. 1.5. Melting temperatures versus lunar depth for Fe, Fe—FeS eutectic and solidus curves for hypothetical lunar mantle (solidus 1) and dry basalt (solidus 2) (from Brett, 1973).

The Moon's Magnetic Field

There is now considerable evidence that lunar surface materials have been magnetized. Most of the current literature holds the general view that part of the magnetization is thermoremanent. The paleofields which magnetized the lunar basalts may have decreased from 1.2 oersteds at 3.95 aeons to 0.33 oersteds at 3.6 aeons (Stephenson et al., 1974). Breccias have also been shown to be magnetic but the origin of their magnetization appears much more complex. The moon's magnetic field was measured directly at several of the Apollo landing sites and the field strengths were found to range from the instrumental threshold of several gamma to more than 300 gamma. Orbital observations of the surface magnetic field indicate that the radial component is typically less than 0.1 gamma. This indicates that very local sources cannot be the origin of these fields, however, regional magnetization is present perhaps associated with regularly magnetized fractured basement rocks (Russell et al., 1973).

The origins of the moon's magnetic field is not known. The most obvious mechanism, an internal dynamo, would require the moon to have a hot core and a high spin rate at the time the lunar surface materials cooled through their Curie temperature. The moon's core, as we have seen, is small and some mechanism must be invoked to slow the moon's spin. If the moon had a planetary spin at birth it is likely that the spin was damped by tidal forces perhaps at the time of capture by the earth. But as discussed in a following section the capture of the moon by the earth would have left a well-defined record in the earth's stratigraphy. Capture, if it occured at all, must have taken place prior to 3.2 aeons ago which appears to be the time of the most recent magnetization. Since the highlands appear to be magnetized a dynamo source field should have operated from nearly the time of accumulation of the moon. The requirements for early differentiation and dynamo action can be met but it is much more difficult to understand how a high lunar spin rate could be maintained after 3.2 aeons unless current views of the earth-moon dynamic history are considerably revised (Sonett and Runcorn, 1973; Sonett, 1974). Since it is difficult to account for a late stage internal dynamo, mare magnetization seems to require a two stage magnetization (Runcorn and Urey, 1973).

It has been proposed that at least part of the lunar field could be provided by impact magnetization (Sonett et al., 1967; Russell et al., 1973). Experimental studies have produced 35 gauss fields with simulated hypervelocity impacts (Srnka, personal communication). Although the dwell times for these events are small, it is likely that magnetization would occur on a regional scale (Sonett, 1974). Impact generated magnetic fields are very-poorly understood and the mechanism has considerable potential, particularly in the magnetization of soil breccias.

Origin of the Moon

The main hypotheses regarding the origin of the moon fall into five broad groupings (1) Binary Planet Hypothesis, (2) Capture Hypothesis, (3) Fission Hypothesis, (4) Precipitation Hypothesis and (5) Sediment Ring Hypothesis. All have good and bad points — none completely explains the body of observational data although in the light of the Apollo missions some hypothesis are more likely than others.

Binary Planet Hypothesis

The binary planet hypothesis argues that the earth-moon system accreted from a common mixture of metal and silicate particles in the same region of the solar nebula. The differences in chemistry and density observed in the two bodies are not easily accounted for in this model in that it requires either that there be inhomogeneities in the solar nebula or that some differentiating process occurs at low temperatures. However, differentiation could occur because of the different physical properties of the metal and silicate particles (Orowan, 1969). During collisions in the solar nebula brittle silicate particles will break up, whereas ductile metal particles could be expected to stick together as they absorb kinetic energy to some extent by plastic deformation. Once metal particles have gathered to form a sufficiently large nucleus, silicate particles could accumulate first by embedding in the ductile metal and later by gravitational attraction. This mechanism produces a cold planetary body with a partially differentiated metallic core already in existence. In terms of the binary planet hypothesis, this simply implies that the earth began nucleating earlier than the moon. The moon is thus less dense and has a smaller metallic core. This also implies that the earth's mantle and the moon have accumulated from the same well-mixed silicate component. Thus basalts formed by partial melting of the earth's mantle and the moon should display a similar compositional range (Ringwood and Essene, 1970). The observed major chemical differences between lunar and terrestrial basalts are not readily explained by this hypothesis. The viability of the hypothesis is further complicated by the fact that the moon does not lie in the plane of the ecliptic as is required if it accreted simultaneously with the earth (Kopal, 1969).

Capture Hypothesis

The capture hypothesis has been the most widely accepted hypothesis in recent times because it can readily explain almost any anomalies in chemical or physical properties (Gerstenkorn, 1955; Alfven, 1963; Mac-Donald, 1964a, 1964b, 1966). This hypothesis argues that the earth and moon accreted in different parts of the solar nebula and that the two planets were brought together at some later date. MacDonald has also proposed a more complex capture hypothesis involving the capture of many small moons which later coalesced to form the present moon. Lunar tidal evolution studies suggested that the moon's orbit was very close to the earth as recently as 2 aeons ago (Gerstenkorn, 1955), and that the moon's surface features were formed at the same time by heating and deformation during the time of close approach (MacDonald, 1964b, Öpik, 1969). The absolute chronology established as a result of the Apollo missions indicates that the major features of the moon are much older (Ringwood and Essene, 1970). Likewise a catastrophic event of this magnitude would appear in the Precambrian record of the earth and would perhaps mark its beginning (Wise, 1969). A normal sedimentary record can be traced back to approximately 3.2 to 3.4 aeons on the earth which suggests that if capture occured it would have to be early in the earth's history. Urey (1962) proposed just that and suggested that capture occured 4.5 aeons ago. However, Urey's hypothesis is based on the idea that the moon is a primitive object with the elements, except for gases, in solar abundances. More recent work by Garz et al. (1969) suggests that the solar Fe/(Mg+Si) ratio is closer to chondritic values than to the moon's chemistry.

The capture of the moon by the earth appears to be an event of low intrinsic probability (Wise, 1969; Ringwood and Essene, 1970). Capture is most likely to occur when the two bodies possess similar orbits which requires that they accrete in the same solar neighborhood presumably from similar parent materials.

Fission Hypothesis

The fission hypothesis was first proposed by Darwin (1880) and has been elaborated by Ringwood (1960), Wise (1963) and Cameron (1966). This hypothesis suggests that the materials now forming the moon were separated from the earth's upper mantle in response to an increasing spin rate driven by segregation of the earth's core. Wise (1969) refined this model by suggesting that the newly fissioned moon may have been baked by the adjacent tidally-heated incandescent earth, resulting in the apparent depletion of volatile alkali elements and enrichment of refractory elements encountered in lunar samples. The appeal of this hypothesis is that the low density of the moon simply relates to the low density of the upper mantle of the earth.

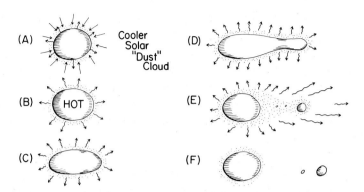

Fig. 1.6. Sequence of events in the fission hypothesis. (from Wise, D. V., *Jour. Geophys. Res., 74:* 6034-6045, 1969; copyrighted by American Geophysical Union).

Figure 1.6 is from Wise (1969) from which the following description of the process of fission is adapted:

(A) The rapidly spinning primordial earth collected from the cooler solar nebula but was heated to incandescence by the 9000-cal g[-1] gravitational energy released by the collection. The surface was at temperatures adequate to volatilize and drive off some of the incoming silicates. Formation of the core (B) released an additional 400 cal g[-1] of gravitational energy and drove the spin rate beyond stability. The last stable figure here is a density stratified Maclaurin spheroid of rotation with axial ratios of approximately 7:12 and a 2.65-hr day. Once stability was reached, the viscous earth passed rapidly through the unstable football-shaped Jacobian ellipsoid (C) and into the bowling-pin-shaped Poincare figure (D), which elongated up to the point of fission. Once fission was achieved, subsidence of the neck back into the earth accelerated the earth slightly, which placed the moon in prograde rotation and initiated its outward migration by tidal frictions. The tidal frictions, amounting to some 800 cal g[-1], were concentrated in the outer shell of the incandescent earth and created a volatilized silicate atmosphere with an extremely high temperature (E). This atmosphere rotating with the planet and subjected to major atmospheric tides from the nearby moon, escaped in an amount equal to a few per cent of the mass of the earth, taking with it most of the angular momentum discrepancy with respect to the present earth—moon system. As the now roasted moon was driven out to a more distant orbit (F), heat generated by tidal friction decreased, and the remains of the silicate atmosphere condensed. Lesser moonlets, formed as smaller droplets from the fission or collected from the debris, were accelerated outward more slowly by tidal friction than the larger moon. Their orbits were perturbed by interaction with the larger moon, forcing the moonlets into more and more elliptical earth orbits until the semimajor axis of the moonlet's orbit 'caught up' with the moon, which was moving outward in its more circular orbit. They impacted into the moon to form the circular mare basins.

Density differences between the near- and farsides, reflecting the denser roots of the bulge in the fission origin, cause the center of mass to be displaced slightly toward the earth with respect to the center of volume. Consequently, the level to which lava can rise hydrostatically is below the surface on the farside but is essentially at the surface on the nearside, resulting in much more complete volcanic flooding of the nearside basins.

Many of the same objections raised for the capture hypothesis can be applied to the fission hypothesis. An event of such magnitude must profoundly affect the geologic record on the earth. Since there is no record of such an event it would have to occur very early in the earth's history. Also, if the moon formed by fission, a much greater similarity could presumably be expected between terrestrial and lunar chemistries. For example, the lunar basalts are much more depleted in siderophile elements than similar basalts on the earth. Further, large scale fractionation appears to have occured very early on the moon suggesting that fission, if it occured, must have been a very early event which seems unlikely. The early history of the moon is discussed in more detail in Chapter 4.

Precipitation Hypothesis

The most recently proposed hypothesis for the moon's origin is the precipitation hypothesis (Ringwood, 1966; Ringwood and Essene, 1970). The hypothesis proposes that the earth accreted over a time interval smaller than 10^7 yr. This would lead to the development of sustained high temperatures, possibly in excess of 1500°C, during the later stages of accretion and to the formation of a massive primitive reducing atmosphere (CO, H_2). These conditions result in the selective evaporation of silicates into the primitive atmosphere while metallic iron continues to accrete on the earth. With the segregation of the core close to the time of the termination of the accretion process the earth's temperature was increased again resulting in the evaporation of some of the silicates in the outer mantle.

Possibly the greatest weakness in this hypothesis is in explaining the energetics of accelerating the atmospheric materials to orbital velocities. Ringwood (1966b) offers the following mechanisms for dissipating the atmosphere —

1. Intense solar radiation as the sun passed through a T-Tauri phase.
2. Interaction of the rapidly spinning terrestrial atmosphere of high-molecular-weight with the solar nebula of low-molecular-weight in which it was immersed.
3. Magnetohydrodynamic coupling resulting in the transfer of angular momentum from the condensed earth to the atmosphere.
4. Rotation instability of the atmosphere due to formation of the earth's core.

Once dissipated, the silicate components of the primitive atmosphere precipitated to form an assembly of planetesimals in orbit around the earth. Further fractionation may also have occured during precipitation because the less volatile components precipitated first and at higher temperatures to form relatively large planetisimals whereas more volatile components precipitated at lower temperatures to form micron-sized particles. The small particles are more likely to be carried away by the viscous drag of escaping gases and lost from the earth-moon system.

Sediment Ring Hypothesis

Ringwood (1970) applied the name "Sediment Ring Hypothesis" to a model proposed by Öpik (1961,1967). This hypothesis compliments both the binary planet hypothesis and the precipitation hypothesis. Öpik's (1961, 1967) model is an attempt to explain the accretion of the moon in terms of lunar cratering and tidal evolution. According to the model, the moon began as a ring of debris within Roche's limit (2.86 earth radii). Clumping of the primordial material resulted in the formation of about six separate rings of debris containing a total of about 1080 fragments. Residual tidal interactions caused a gradual outward migration of the debris rings. As each ring of debris passes beyond Roche's limit it rapidly accreted to form a moonlet 1/6 of the moon's mass and 1910 km in diameter and developed temperatures up to 850°K. Gradually the separation between the moonlets decreased until partial tidal disruption followed by merger occured. The final merger occured at 4.5 to 4.8 earth radii. Small fragments produced during the merger collided with the moon to produce the dense pre-mare cratering in the highlands.

References

Alfven, H., 1963. The early history of the moon and the earth. *Icarus, 1:* 357-363.

Baldwin, R. B., 1971. The question of isostasy on the moon. *Phys. Earth Planet. Interiors, 4:* 167-179.

Brett, R., 1972. Sulfur and the ancient lunar magnetic field. *Trans Am. Geophys. Union, 53:* 723.

Brett, R., 1972. A lunar core of Fe-Ni-S. *Geochim. Cosmochim. Acta, 37:* 165-170.

Cameron, A. G. W., 1966. Planetary atmosphere and the origin of the moon. In: B. S. Marsden and A. G. W. Cameron (Editors), *The Earth-Moon System,* Plenum Press, N. Y., 234 pp.

Church, S. E., Bansal, B. M., and Wiesman, H., 1972. The distribution of K, Ti, Zr, U, and Hf in Apollo 14 and 15 materials. In: *The Apollo 15 Lunar Samples,* The Lunar Science Institute, Houston, Texas, 210-213.

Darwin, G. H., 1880, On the secular charge in elements of the orbit of a satellite revolving about a tidally distorted planet. *Phil. Trans. Roy. Soc. Lon., 171:* 713-891.

Duba, A. and Ringwood, A. E., 1973. Electrical conductivity, internal temperatures and thermal evolution of the moon. *The Moon, 7:* 356-376.

Duncan, A. R., Ahrens, L. H., Erlank, A. J., Willis, J. P., and Gurney, J. J., 1973. Composition and interrelationships of some Apollo 16 samples. *Lunar Science—IV,* The Lunar Science Inst., Houston, Texas, 190-192.

Erlank, A. J., Willis, J. P., Ahrens, L. H., Gurney, J. J. and McCarthy, T. S., 1972. Inter-element relationships between the moon and stony meteorites, with particular reference to some refractory elements. *Lunar Science—III,* The Lunar Science Inst., Houston, Texas, 239-241.

Garz, T., Kock, M., Richter, J., Baschek, B., Holweger, H. and Unsold, A., 1969. Abundances of iron and some other elements in the sun and in meteorites. *Nature, 223:* 1254.

Gerstenkorn, M., 1955. Uber Gezeitenreibung bei Zweikorpenproblem. *J. Astrophys., 26:* 245-279.

Gilbert, G. K., 1893. The Moon's face, a study of the origin of its features. *Philos, Soc. Washington Bull., 12:* 241-292.

Green, D. H., Ringwood, A. E., Ware, N. G., Hibberson, W. O., Major, A., and Kiss, E., 1971. Experimental petrology and petrogenesis of Apollo 12 basalts. *Proc. Second Lunar Sci. Conf., Suppl. 2, Geochim. Cosmochim. Acta, 1:* 601-615.

Hanks, T. C. and Anderson, D. L., 1969. The early thermal history of the earth. *Phys. Earth Planet. Interiors, 2:* 19-29.

Hansen, M. and Anderko, K., 1958. *Constitution of the binary alloys.* McGraw-Hill, N. Y., 2nd Edit., 720 pp.

Kopal, Z., 1969. *The Moon.* Reidel, Dordrecht, 525 pp.

Latham, G. V., Ewing, M., Press, F., Sutton, G., Dorman, J., Nakamura, Y., Toksoz, N., Lammlein, D., and Duennebier, F., 1972. Passive seismic experiment. *Apollo 16 Preliminary Science Report,* NASA SP-315, 9-1 to 9-29.

MacDonald, G. J. F., 1966. Origin of the moon: Dynamical consideration, In: B. S. Marsden and A. G. W. Cameron (Editors), *The Earth-Moon Systems,* Plenum Press, N. Y., 165-209.

MacDonald, G. J. F., 1964a. Earth and Moon: Past and future. *Science, 145:* 881-890.

MacDonald, G. J. F., 1964b. Tidal friction. *Rev. Geophys., 2:* 467-541.

McConnell, R. K. and Gast, P. W., 1972. Lunar thermal history revisited. *The Moon, 5:* 41-51.

Nakamura, Y., Latham, G., Lammlein, D., Ewing, M., Duennebier, F. and Dorman, J., 1974. Deep lunar interior inferred from recent seismic data: *Geophys. Res. Letters, 1:* 137-140.

Öpik, E. J., 1961. Tidal deformations and the origin of the moon. *Astrophys. Jour., 66:* 60-67.

Öpik, E. J., 1967. Evolution of the moon's surface I. *Irish Astron. Jour., 8:* 38-52.

Öpik, E. J., 1969. The moon's surface. *Ann. Rev. Astron. Astrophys., 7:* 473-526.

Orowan, E., 1969. Density of the moon and nucleation of planets. *Nature, 222:* 867.

Ringwood, A. E., 1960. Some aspects of the thermal evolution of the earth. *Geochim. Cosmochim. Acta, 20:* 241-259.

Ringwood, A. E., 1970a. Origin of the moon: the precipitation hypothesis. *Earth Planet. Sci. Lett., 8:* 131-140.

Ringwood, A. E., 1970b. Petrogenesis of Apollo 11 basalts and implication for lunar origin. *Jour. Geophys. Res., 75:* 6453-6479.

Ringwood, A. E. and Essene, E., 1970. Petrogenesis of Apollo 11 basalts, internal constitution and origin of the moon. *Proc. Apollo 11 Lunar Sci. Conf., Suppl. 1, Geochim. Cosmochim. Acta, 1:* 769-799.

Runcorn, S. K. and Urey, H. C., 1973. A new theory of lunar magnetism. *Science, 180:* 636-638.

Russell, C. T., Coleman, P. J., Lichtenstein, B. R., Schubert, G. and Sharp. A. R., 1973. Subsatellite measurements of the lunar magnetic field. *Proc. Fourth Lunar Sci. Conf., Suppl. 4, Geochim. Cosmochim. Acta, 3:* 2833-2845.

Shaler, N. S., 1903. A comparison of features of the earth and the moon. *Smithsonian Contr. Knowledge*, 34, pt. 1, 130 pp.

Sonett, C. P., Colburn, D. S., Dyal, P., Parking, C. W., Smith, B. F., Schumbert, G. and Schwartz, K., 1971. Lunar electrical conductivity profile. *Nature, 230:* 359-362.

Sonett, C. P., 1974. Evidence of a primodial solar wind. *Proc. Third Solar Wind Conf.,* Inst. of Geophysics and Planetary Physics, Univ. of Calif., Los Angeles, 36-57.

Sonett, C. P. and Runcorn, S. K., 1973. Electromagnetic evidence concerning the lunar interior and its evolution. *The Moon, 8:* 308-334.

Stephenson, H., Collinson, D. W., and Runcorn, S. K., 1974. Lunar magnetic field paleointensity determinations on Apollo 11, 16 and 17 rocks. *Proc. Fifth Lunar Sci. Conf., Suppl. 5. Geochim. Cosmochim. Acta, 3:* 2859-2871.

Sterrett, K. F., Klement, W., and Kennedy, G. D., 1965. Effect of pressure on the melting of iron. *Jour. Geophys. Res., 70:* 1979-1984.

Stuart-Alexander. D., and Howard, D. A., 1970. Lunar maria and circular basins – a review. *Icarus, 12:* 440-456.

Taylor, S. R., Gorton, M. P., Muir, P., Nance, W., Rudowski, R. and Ware, N. 1973. Lunar highland composition: Apennine Front. *Proc. Fourth Lunar Sci. Conf., Suppl. 4, Geochim. Cosmochim. Acta, 2:* 1445-1459.

Taylor, S. R. and Jakes, P., 1974. The geochemical evolution of the moon. *Proc. Fifth Lunar Sci. Conf., Suppl. 5, Geochim. Cosmochim. Acta, 2:* 1287-1305.

Toksoz, M. N., Solomon, S. C., Minear, J. W., and Johnston, D. H., 1972. The Apollo 15 lunar heat flow measurement. *Lunar Science—III,* The Lunar Science Inst., Houston, Texas, 475-477.

Urey, H. C., 1962. Origin and history of the moon. In: Z. Kopal (Editor), *Physics and Astronomy of the Moon.* Academic Press, London, 481-523.

Walker, D., Longhi, J., Stolper, E., Grove, T., Hays, J. F., 1974. Experimental petrology and origin of titaniferous lunar basalts. *Lunar Science—V,* The Lunar Science Inst., Houston, Texas, 814-816.

Wanke, H., Baddenhausen, H., Dreibus, G., Jagoutz, E., Kruse, H., Palme, H., Spettel, B., and Teschke, F., 1973. Multielement analyses of Apollo 15, 16 and 17 samples and the bulk composition of the moon. *Proc. Fourth Lunar Sci. Conf., Suppl. 4, Geochim. Cosmochim. Acta, 2:* 1461-1481.

Wise, D. U., 1969. Origin of the moon from the earth; some new mechanisms and comparisons. *Jour. Geophys. Res., 74:* 6034-6045.

Wood, J., 1972. Asymmetry of the moon, *Lunar Science—IV,* The Lunar Science Inst., Houston, Texas, 790-792.

Wood, J. A., 1972. Thermal history and early magnetism in the moon. *Icarus, 16:* 229-240.

CHAPTER 2

Energy at the Lunar Surface

Introduction

By its nature a sediment or sedimentary rock represents expended energy. It is the product, to a large degree, of its physical environment. On the earth's surface the physical environment of the planet is dominated by the atmosphere and the hydrosphere both acting as intermediaries in the transfer of solar energy for the breakdown and dispersal of sedimentary materials. The moon by contrast very obviously lacks both an atmosphere and a hydrosphere. This means that solar energy, which is by far the most important energy source for the formation of terrestrial sediments, is almost completely ineffective in this role on the lunar surface. However, the absence of an atmosphere and hydrosphere on the moon leaves the planetary surface bared to the interplanetary environment and kinetic energy in the form of impacting meteoroids. It also leaves other processes such as volcanism, mass movement and electrostatic processes unimpeded on the lunar surface.

In this chapter the various potential energy sources are evaluated in terms of their abilities to erode or transport detrital materials on the lunar surface. The significance of some energy sources, such as meteoroid flux, is obvious while others, such as the potential value of solar radiation, are much debated.

The Meteoroid Flux

The surface of the moon is very obviously cratered and whether one subscribes to an impact or volcanic origin for the majority of the craters it is clear that meteoroid impact has played a major role in sculpturing the lunar surface.

The meteoroid cloud has a total mass of 2.5×10^{19} g within 3.5 AU of the sun and within an inclination of $i < 20°$ of the ecliptic (Whipple, 1967). This represents a total volume of 2.25×10^{41} cm^3. This cloud produces a continuous rain of hypervelocity particles on the lunar surface. The rate at which energy is applied to the lunar surface is small in terrestrial terms but

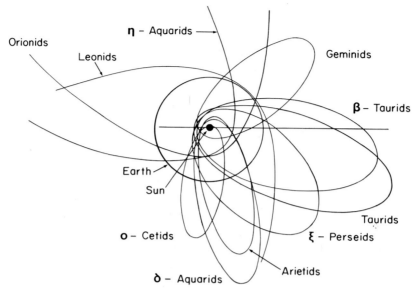

Fig. 2.1. Selected orbits of meteoroid streams.

with the long periods of geologic time available it is a highly effective mechanism for erosion and transportation on the lunar surface.

Distribution of Meteoroids in Space

Two types of meteor populations can be distinguished according to their spatial distribution; shower meteors and sporadic meteors (Dohnanyi, 1970). Shower meteors move in highly correlated orbits and are distributed into the volume of an elliptical doughnut with the sun in one of its foci (Fig. 2.1). Close correlation between the orbit of a meteor shower and that of a known comet has been established for a number of meteor showers, and it is believed that meteor showers originate from the partial disruption of comets (Whipple, 1963). Such disruptions occur when the comet is subjected to the thermal forces caused by the sun's radiation (Whipple, 1963).

Sporadic meteoroids, on the other hand, move in random orbits and 'fill up' the interplanetary space. The populations of sporadic and shower meteoroids are believed to be closely related (Jacchia, 1963). Sporadic meteoroids may be shower meteoroids that have gone astray under the perturbing influence of the planets (Plavec, 1956).

Two gravitational dispersal processes can be distinguished for meteor streams. One dispersal effect arises when particles are ejected from the comet near perihelion with a relatively small and limited ejection velocity (Whipple,

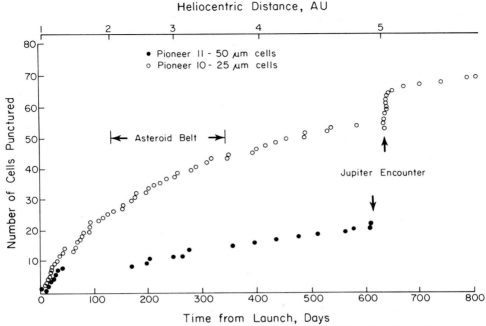

Fig. 2.2. Time history of meteoroid penetrations detected by Pioneer 10 and 11 (from Humes *et al.*, 1975, copyright 1975 by The American Association for the Advancement of Science).

1950, 1951). The result is a distribution of orbital elements for the particles around a mean value similar to the orbital elements of the parent comet. The particles thus assume a spatial distribution of a cloud whose dimension is defined by the spread in the orbital elements of the particles. However, with time the distribution in the periods of the particles causes a gradual dispersal of the cloud along an orbit similar to that of the parent comet until the shower particles are confined into a doughnut-like region of space whose shape is similar to the orbit of the parent comet.

The second perturbing force acting on the meteor shower is the influence of planetary encounters. When a particle or group of particles encounters a planet, they scatter into orbits that are quite different from their initial orbits. The magnitude of the effect depends on the proximity of the encounter. The long-range statistical effect is that the particles will assume a distribution of orbits having an average inclination/eccentricity ratio similar to that of the sporadic meteors (McCrosky and Posen, 1961; ·Öpik, 1966).

The most direct evidence for the distribution of meteoroids between the earth and Jupiter has come from impact detectors on the Pioneer 10 and 11 space probes (Humes *et al.*, 1975) (Fig. 2.2). The distribution of meteoroids with masses less than 10^{-8} g is relatively constant out to a distance of almost 6 A.U. There is a slight increase in meteoroids between 1 and 1.15 A.U. and a complete lack of meteoroids, or at least meteoroid penetrations, between 1.15 and 2.3 A.U. There was no evidence of an increase of the meteroid density in the asteroid belt which tends to refute any therory that the asteroids are the source of small meteoroids. A high concentration of meteoroids was encountered in the vicinity of Jupiter and is believed to be due to gravitational focusing by this planet.

Information on the population of meteoritic objects beyond the orbit of Jupiter is limited. However, estimates of the spatial distribution of radio meteors has given some insight into the overall distribution of meteoroids in the solar system (Southworth, 1967b). The fraction of time each meteor spends in given regions of space was calculated and in combination with weighting factors yields an estimate for the spatial distribution (Fig. 2.3). The distribution is assumed to be symmetric about an axis through the ecliptic poles; there is a broad maximum in the ecliptic and a minimum at the poles. The distribution of radio meteors is flatter than the one based on the photometry of the zodiacal cloud (Fig. 2.3). Detection of fast meteors (those in highly eccentric orbits with large semimajor axes) is strongly favored by radar and, consequently, the emperical distribution of fast meteors will be greatly exaggerated.

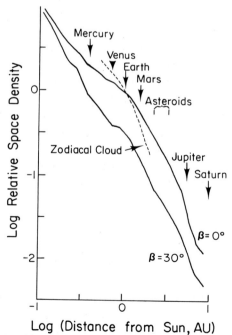

Fig. 2.3. Relative space density of radio meteors and of zodiacal particles as a function of their distance from the sun, and ecliptic latitude β (from Dohnanyi, 1972).

Measuring the Meteoroid Flux

A large volume of data is available for the mass distribution of meteoroids. The data have been derived by a variety of methods, some earth based, others using spacecraft and more recently lunar rocks. Consequently, the results and interpretations are frequently contradictory and a wide assortment of meteoritic models and fluxes has been suggested by various workers (Dohnanyi, 1970, 1972; Gault, *et al.*, 1972; Kerridge, 1970; Soberman, 1971, McDonnell, 1970).

The mass distribution of micrometeoroids (particles smaller than 10^{-6} g) has been determined by a variety of methods. The first observation of micrometeoroids in the solar system was obtained by ground measurements. Observations of the solar K-corona (Allen, 1946) and the zodiacal light at greater elongations (Van de Hulst, 1947) suggested that the two phenomena arise from the scattering of sunlight by cosmic dust. At the antisolar point, the brightness of the zodiacal cloud increases and forms a diffuse area of about $(10^0)^2$ called the gegenschein. The gegenschein appears to result from the backscatter of the sun's light by the zodiacal cloud (Siedentopf, 1955) although there are other views (Roosen, 1970) which suggest that the gegenschein is due to a particle concentration in the asteroidal belt or beyond.

With the coming of the space program a number of different detectors based on particle penetrations were flown on spacecraft (Fechtig, 1971 and Dohnanyi, 1972 give reviews). These detectors have the advantage of making direct measurements of the properties of individual micrometeoroids and therefore provide much more detailed information than can be obtained optically from the zodiacal cloud. The results of penetration measurements are of considerable importance because they provide the only direct measurements of micrometeoroids. Pressurized can detectors were used on some Explorer and Lunar Orbiter satellites. The Ariel satellites used a thin aluminum foil detector which permitted the passage of light through holes produced by micrometeoroid impacts (Jennison *et al.*, 1967). Capacitor detectors were used on some Pegasus and Pioneer satellites. These detectors are momentarily discharged by impact ionization as the micrometeoroid penetrates the detector. Other Pioneer satellites had two parallel detecting surfaces which allowed the determination of velocity and direction of motion and allowed the calculation of orbital parameters. These Pioneer detectors were also backed with a microphone which detects the momentum of the events.

A certain amount of information concerning the micrometeoroid flux is available from more direct evidence. For example, possible meteoritic particles have been collected by rockets and balloons in the upper atmosphere (Farlow and Ferry, 1971). Information is also available from deep ocean sediments and particulate material in polar ice. Information from these sources is much more difficult to interpret as it is never clear as to what fraction of the particulates is extraterrestrial. More promising perhaps is the use of trace element chemistry to estimate the extraterrestrial contribution to deep ocean sediments (Baker and Anders, 1968). Similarly, several flux estimates are available from trace element studies of lunar soils (Ganapathy et al., 1970a, 1970b; Keays et al., 1970; Hinners, 1971). These estimates are discussed in more detail in a following section dealing with the availability of meteoritic energy at the lunar surface; they provide little information on the mass distribution of the meteoritic particles.

The surfaces of most lunar rocks are densely pitted by micrometeoroid impacts. These pits, most of which are glass lined, range in size up to the order of millimeters and provide very direct data on the mass distribution of micrometeoroids (Hörz et al., 1971; Hartung et al., 1971, 1972; Gault et al., 1972). At least 95 per cent of the observed pits were the product of micrometeoroids impacting at velocities in excess of 10 km sec^{-1} (Hörz et al., 1971). Some rocks have an equilibrium distribution of craters (Marcus, 1970; Gault, 1970) but younger glass surfaces of impact melt are not saturated. Similar but smaller estimates of the present lunar flux were obtained by examining craters on the surface of parts of the Surveyor III spacecraft returned from the moon by Apollo 12 (Cour-Palais et al., 1971).

A variety of methods of greater and lesser reliability are available for looking at the small end of the meteoroid mass distribution. In some ways it is easier to study larger meteoroids simply because they are large, for example, they frequently reach the earth's surface so we know much more about their chemistry than we do about micrometeoroids. However, arriving at estimates of their pre-atmospheric mass distribution is considerably more complex largely because the observational methods are indirect and require difficult calibrations.

Meteoroids entering the earth's atmosphere ionize the air molecules in their path and create an ionized trail. Intensely ionized trails of long duration may be detected by radio equipment and are then called radio meteors. Determination of the distribution of radio meteors as a function of mass and orbital elements is difficult because of selection effects and because of the uncertain relation between the mass of a radio meteor and its observable dynamic parameters. Because of selection effects, radar misses the fastest and slowest meteors. Fast meteors leave ionized trails at great heights (above 100 km) and the trail charge density may, at these altitudes, diffuse into the background before detection. Slow meteors do not leave a sufficiently high electron line density that can be detected. Consequently, radar favors the detection of meteors having an earth entry speed of about 40 km s^{-1}; selection effects inhibit the detection of meteors with appreciably different velocities.

Luminous trails produced by meteoroids entering the earth's atmosphere are bright enough to be photographed. These trails are produced by meteoroids with masses on the order of milligrams or greater (Dohnanyi, 1972). As with radio meteors the greatest difficulty in determining meteor masses relates to calibration problems and strong selection effects. The luminous intensity of a meteor is proportional to the third power of its velocity over certain velocity ranges. Consequently there is a strong preference for fast meteors to be detected because of the higher relative luminosity. Because of the uncertainties in calibration photometric data provide an order of magnitude approximation of the mass distribution (Jacchia and Whipple, 1961; Hawkins and Southworth, 1958; McCrosky and Posen, 1961; Dohnanyi 1972).

Possibly one of the most promising means of determining the flux of larger meteoroids is the use of the seismic stations established on the moon during the Apollo missions (Latham et al., 1969). The advantage of the method is that essentially the whole moon is being used as a detector so that it is possible to obtain large samples of relatively rare events. The method has not yet reached its full potential, once again because of calibration difficulties. Estimates of the flux using the short period component of the seismic data compared favourably with some earth-based observations, but it implies a considerably higher flux than follows from analyses of long-period lunar seismic data (Duennebier and Sutton, 1974; Latham et al., 1972). The discrepancy is believed to lie in the uncertainty of coupling of impact energy to seismic waves.

The asteroid belt lies in the region between the planets Mars and Jupiter at about 2.5 to 3.0 AU from the sun. Several thousand asteroids occur in this diffuse doughnut shaped region and are readily observed optically (Figure 2.4). The asteroids are important to the present discussion for two reasons. First, they provide us with information on the mass distribution of the larger meteoritic bodies (m < 10^{13} g) and second, they figure prominently in many models of the origin and evolution of the meteoroid flux.

Attempts have been made to use lunar and terrestrial crater counts as a means of measuring time and correlating cratered surfaces on an interplanetary basis (Shoemaker et al., 1962). Counts of large craters in the central United States and on the Canadian shield were used to determine cratering rates and hence the flux of large bodies. Numerous similar crater

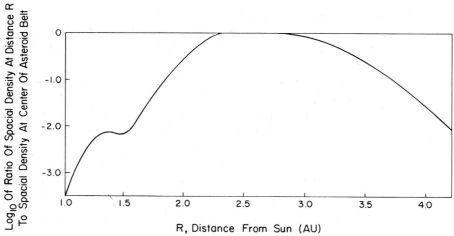

Fig. 2.4. Radial distribution of asteroids (from Dohnanyi, 1972).

counts were made of surfaces of different age on the moon. With absolute ages now available for some lunar surfaces as a result of the Apollo program the data will undoubtably lead to a considerable refinement of the large-body flux (Table 2.1).

The Mass-Frequency Distribution

The array of available flux data is both complex and confusing. In almost all cases complex calibrations and corrections are required to convert

TABLE 2.1

Determinations of post-mare terrestrial and lunar cratering rate (from Dohnanyi, 1972).

Method	Number of craters of diameter >1 km $(km^2\ 10^9\ years)$	
	Earth	Moon
Meteorite and asteroid observations	21×10^{-4}	12×10^{-4}
	5 to 23×10^{-4}	2 to 9×10^{-4}
	110×10^{-4}	45×10^{-4}
Astrobleme counts in central U.S.A.	1 to 10×10^{-4}	0.5 to 4×10^{-4}
Crater counts in Canadian shield	1 to 15×10^{-4}	0.4 to 6×10^{-4}
Best estimate	12×10^{-4}	5×10^{-4}

observational data to number-frequency equilvalents. In figures 2.5 and 2.6 most of the available data are plottted and it can be seen that in some cases there are differences of several orders of magnitude. A large number of writers have reviewed the literature dealing with the statistics of meteoroid masses (Vedder, 1966; Whipple, 1967; Bandermann, 1969; Bandermann and Singer, 1969; Kerridge, 1970; Soberman, 1971; McDonnell. 1970). Few reviewers have, however, examined the data critically and attempted to establish a best estimate of the flux. The most recent attempt to this end was provided by Gault *et al.* (1972).

All of the micrometeoroid data selected by Gault *et al.* (1972) were required to fulfill three criteria (Kerridge, 1970). First, the original data must have come from an experiment that gave a positive and unambiguous signal of an event. Second, it was required that the experiment must have recorded sufficient micrometeoroid events to provide a statistically significant result. Third, the response and sensitivity of the sensors must have been calibrated using hypervelocity impact facilities in order to provide, as far as possible, a simulation of micrometeoroid events. Only 10 data points from

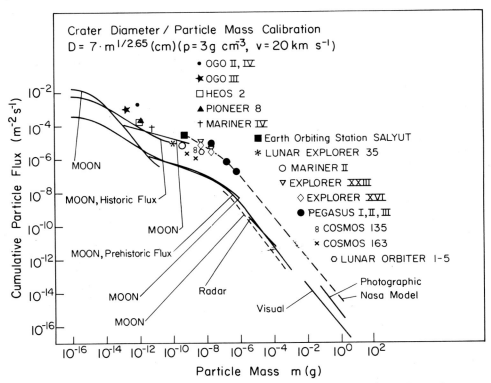

Fig. 2.5. Comparison of lunar and satellite micrometeoroid flux data (after Hörz *et al.*, 1973).

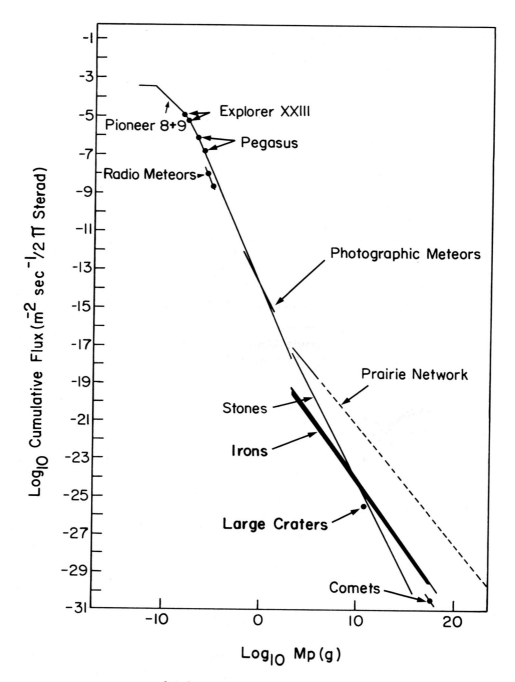

Fig. 2.6. Cumulative flux ($m^{-2}sec^{-1}/2\ \pi$ sterad) of meteoroids and related objects into Earth's atmosphere having a mass of $M_p(g)$ or greater (after Dohnanyi, 1972).

experiments meet all of the requirements (Fig. 2.7).

For larger particles with masses between 10^{-6} and 1g Gault *et al.* (1972) found that recent analyses of photographic meteor data (Naumann, 1966, Lindblad, 1967; Erickson, 1968; Dalton, 1969; Dohnanyi, 1970, 1972) agree closely with the earlier NASA model (Cour-Palais, 1969). After modification for the lunar environment the NASA model was adopted. For masses greater than 1 g Hawkins (1963) earlier model was adopted. The Hawkins model is in disagreement with the model based on long-period seismic data (Latham *et al.*, 1971). However, as pointed out previously, there appear to be serious calibration problems associated with the seismic model.

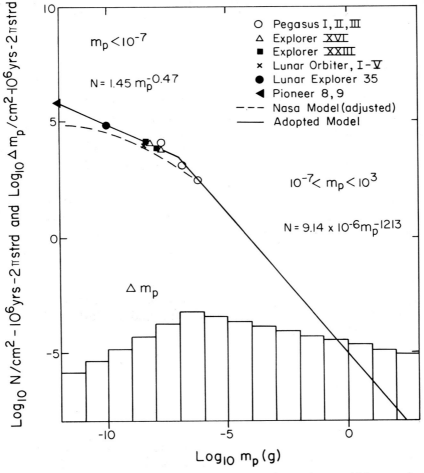

Fig. 2.7. The micrometeoroid flux measurements from spacecraft experiments which were selected to define the mass-flux distribution. Also shown is the incremental mass flux contained within each decade of m_p (after Gault *et al.*, 1972, *Proc. 3rd Lunar Sci. Conf., Suppl. 3, Vol. 3*, MIT/Pergamon).

The mass distribution for the lunar meteoritic flux of N particles (per cm^{-2} 10^6 yr^{-1} 2π $strd^{-1}$) of mass M_p (g) according to the model is:

$$N = 1.45\, M_p^{\,0.47} \qquad 10^{-13} \leq M_p \leq 10^{-7} \qquad (2.1)$$
$$N = 9.14 \times 10^{-6}\, M_p^{\,1.213} \qquad 10^{-7} \leq M_p \leq 10^3 \qquad (2.2)$$

for particles with a density of 1 g cm^{-3} impacting with a root-mean-square velocity of 20 km s^{-1}. The use of two exponential expressions with a resultant discontinuity is artificial but was introduced for simplicity. Also shown in Figure 2.7 is the incremental mass flux in each decade of particle mass, which shows quite clearly that most of the mass of materials impacting on the lunar surface consists of particles of the order of 10^{-6} g.

Velocity Distribution

The velocity distribution of meteoroids is bounded on the lower side by the escape velocity of the earth and on the upper side by the escape velocity of the solar system (Figure 2.8). The modal velocity of the distribution is approximately 20 km s^{-1}. This last figure is generally accepted as the root-mean-square velocity for meteoroids (Gault et al., 1972). Particles impacting the moon may have velocities as low as 2.4 km s^{-1} (the lunar escape velocity) but the mean is expected to be closer to 20 km s^{-1} because of the relatively lower effective cross section of the moon for slower particles. The velocity-frequency distribution is noticeably skewed towards the higher velocities with a secondary mode near the escape velocity for the solar system. The secondary mode undoubtably reflects meteoroids with a cometary origin. The distribution in Figure 2.8 is based on radio and photo meteors, however, the curve is generally consistent with inflight velocity measurements of micron-size particles (Berg and Gerloff, 1970; Berg and Richardson, 1971).

Physical Properties

Density

A search of the literature produces a large number of values of the mean density of the meteoroid flux. Most, however, are just estimates and have frequently been made to satisfy the needs of the particular study. A considerable body of information is also available for terrestrial falls.

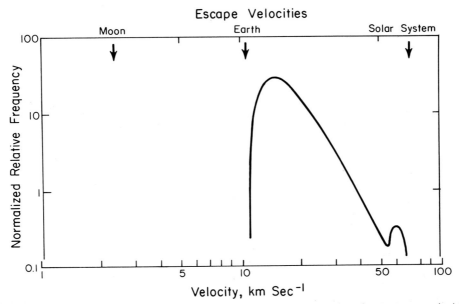

Fig. 2.8. Velocity frequency distribution of meteoroids in relation to the relevant escape velocities (after Hartung *et al.*, 1972, *Proc. 3rd Lunar Sci. Conf., Suppl. 3, Vol. 3*, MIT/Pergamon).

Unfortunately, the survival of a meteorite during entry into the earth's atmosphere as well as its chances of being found is very dependent upon its composition. For example, iron and stony meteorites are much more likely to survive than carbonaceous meteorites or meteorites with a high gas content. Because density is also composition dependent it is difficult to assess the statistics of this information. Information from terrestrial finds does, however, suggest that the iron meteorites provide a useful figure for an upper density limit approximately 8 g cm^{-3}. Much of the experimental work suggested that micrometeoroids were "fluffy" and thus of very low density (eg. <1 g cm^{-3}) (Soberman, 1971). The geometry of microcraters on lunar rocks has been analysed and compared with experimental data to extract information on particle density. The ratio of the maximum crater depth below the original uncratered surface to the mean diameter of the pit rim is a function of particle density and velocity or impact (Fig. 2.9). The data are completely inconsistent with micrometeoroid densities being less than unity and suggest that very few particles have the density of iron (Hörz *et al.*, 1973). If it is assumed that the root-mean-square velocity of meteoroids is 20 km s^{-1} most meteoroids must have densities between 2 and 4 g cm^{-3}.

Approximately 10 per cent of the microcrater population on lunar rock surfaces appear to be different. Some of these microcraters have no glass lining which has been interpreted by some to suggest that they are produced

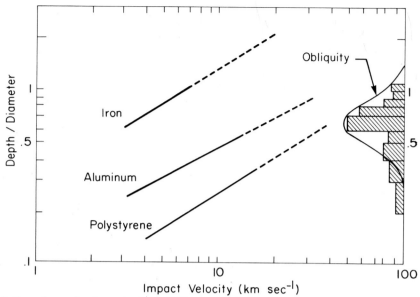

Fig. 2.9. Experimentally determined depth/diameter ratios using projectiles with densities from 1 to 7.3 g cm^{-3} and impact velocities from 3 to 13 km sec^{-1}. The inserted histogram on the lunar depth/diameters is based on 70 craters; an empirical curve accounting for the oblique impact is indicated (from Hörz et al., 1973).

by low velocity impacts — possibly secondary events. However, there is evidence to suggest that the craters were initially glass lined and the glass has been spalled off. Possibly more important is another rare type of crater which has been called a multiple pit crater (Hörz et al., 1973). These microcraters are produced by aggregate structures with a low density and non-homogenous mass distribution.

Shape

Direct observation of the shape of meteorites provides little useful information. Mason (1962) presents photographs of a variety of meteorite shapes, however, most were determined by ablation during entry into earth's atmosphere and later by weathering on the ground. However, crater symmetry is determined to a large extent by the shape of the projectile and its angle of incidence (Mandeville and Vedder, 1971; Kerridge and Vedder, 1972; Mandeville and Vedder, 1973). Most noncircular microcraters on lunar rocks are elongate and shallow which indicates that they were produced by an oblique impact rather that by irregularly shaped micrometeoroids (Hörz et al., 1973). Thus highly nonspherical micrometeoroids such as rods or platelets are very unlikely. If the particles are modeled as prolate elipsoids the ratio a/b is less than 2. As mentioned in the preceding section, multiple pit craters suggest that a small number of micrometeoroids may have been complex aggregate particles.

Chemical Composition

The chemistry of terrestrial meteorite finds is well documented in numerous publications (eg. Mason, 1971) and is beyond the scope of the present book. Because of the atmospheric entry conditions, weathering at the earth's surface and the unknown statistical parameters connected with the finding of meteorites, an estimate of the bulk chemistry of the flux on the basis of terrestrial finds is at best difficult.

Three types of meteoritic material could be expected on the moon (Anders et al., 1973):

1. Micrometeoroids and small meteoroids in the soil.
2. Crater forming bodies in ray material and other ejecta.
3. Planetesimals from the early intense bombardment of the moon in ancient breccias and highland soils.

Few meteoritic particles survive a hypervelocity impact. Consequently a study of the chemistry of meteoritic materials on the moon must be based on the bulk properties of lunar rocks and soils in relation to elements which are distinctive of meteoritic materials. Originally, it appeared the siderophile elements (Ir, Au, Re, Ni, etc.) would be the most reliable indicators of meteoritic material (Anders et al., 1973). Because they concentrate in metal phases during planetary melting, they are strongly depleted on the surfaces of differentiated planets (e.g. by a factor of 10^{-4} on Earth). Accordingly, they had been used as indicators of meteoritic material in oceanic sediments and polar ice (Barker and Anders, 1968; Hanappe et al., 1968). However, a number of volatile elements (Ag, Bi, Br, Cd, Ge, Pb, Sb, Se, Te, Zn) turned out to be so strongly depleted on the lunar surface that it became possible to use some of them as subsidiary indicator of meteoritic matter.

Micrometeoroid Composition

All lunar soils are enriched in meteoritic indicators—elements compared to their crystalline source rocks. However, because of lateral mixing not all of this enrichment can be attributed to meteoritic materials. Most soils contain at least small amounts of exotic rock types which are not found among the large rock fragments collected at the site. Most important are the alkali rich (granitic and noritic) rocks which result in the soils being enriched

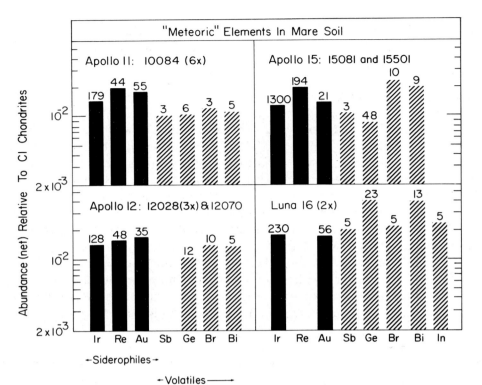

Fig. 2.10. All mare soils studied to date are enriched in 'meteoritic' elements, relative to crystalline rocks. Net meteoritic component is obtained by subtracting an indigenous lunar contribution, estimated from crystalline rocks. Numbers above histogram indicate signal-to-noise ratio (i.e. ratio of net component to correction). Abundance pattern is flat, with siderophiles and volatiles almost equally abundant. Apparently the meteoritic component has a primitive composition (cf. Fig. 2.11) (from Anders *et al.*, 1973).

in alkalis, uranium and thorium. After correcting for this enrichment Anders *et al.* (1973) found that the soil at the four sites examined all show essentially the same picture, with siderophiles enriched to 1.5-2 per cent Cl chondrite equivalent and volatiles enriched to a similar if more variable extent. In figure 2.10 the data are normalized to Cl chondrites to permit characterization of the meteoritic component. The abundance pattern is flat with siderophile and volatile elements almost equally abundant. This pattern rules out all fractionated meteoroid classes when compared to the right side of Figure 2.11. Ordinary chondrites, E5-6 chondrites, irons, and stony irons are too deficient in volatiles (especially Bi) relative to siderophiles, whereas achondrites are too low in siderophiles. The attention is then focused on primitive meteorites on the left side of the figure in which all siderophiles and volatiles occur in comparable abundances. A weak trend in the Apollo data suggests a slight depletion of volatiles that may be due to a small

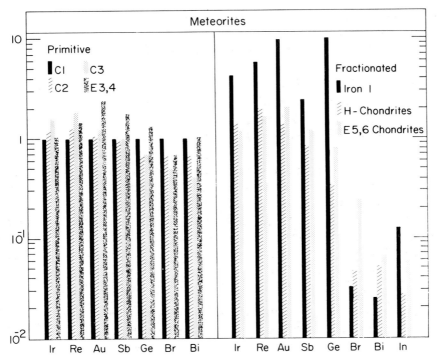

Fig. 2.11. Primitive meteorites (left) contain siderophiles and volatiles in comparable abundance. Fractionated meteorites (right) are depleted in volatiles (from Anders *et al.*, 1973).

admixture of fractionated material (Anders *et al.*, 1973). However, the dominant material appears to be of Cl composition.

Several workers have taken a different and somewhat less rewarding approach to the problem of micrometeoroid composition (Chao *et al.*, 1970; Bloch *et al.*, 1971; Schneider *et al.*, 1973). Glasses lining the pits of microcraters were analyzed in an attempt to detect exotic components which could be identified as part of the original projectile. In general the results of these studies have been negative. This has been interpreted to indicate that most micrometeoroids are in large part silicates which is in agreement with the density estimates given by Hörz *et al.* (1973) and with the findings of Anders *et al.* (1973).

Planetesimals

The older soils exposed in the lunar highlands have a different and more complex diagnostic meteoritic element pattern than the mare soils (Anders *et al.*, 1973) which suggests that the ancient meteoritic component is compositionally different from the micrometeoroid component. The greatest difficulty in dealing with the highland soils is estimating the correction for indigenous materials.

Siderophile elements, especially Au, are more abundant in highland than in mare soils (approximately 2-4 per cent of the bulk soil) (Fig. 2.12). Four volatiles (Sb, Ge, Se, Bi) occur as in the mare soils at the 1-2 per cent level but Zn and Ag are markedly higher at 5-9 per cent Cl chondrite equivalent. The high Zn and Ag values may relate to a rare indigenous rock type but most of the remaining differences are due to an ancient meteoritic component.

The ancient component is much more intimately associated with its host rock than the recent micrometeoroid component. The latter is located mainly on grain surfaces, as shown by its ready acid leachability (Silver, 1970; Laul *et al.*, 1971) and higher abundance in smaller grain size fractions (Ganapathy *et al.*, 1970b). This is expected for material vaporized and recondensed on impact. The ancient component, on the other hand, is distributed throughout the interior of massive breccias. This difference may reflect a major difference in size of the incoming objects. Breccia formation

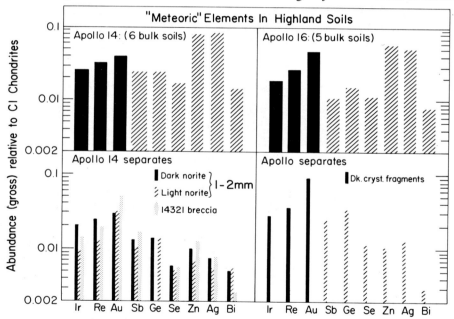

Fig. 2.12. Bulk highland soils (top) show a less regular pattern than mare soils, largely due to an 'ancient' meteoritic component. This component is seen most clearly in breccias and coarse soil separates which contain no recent micrometeoroid contribution (bottom). It has a distinctly fractionated pattern, with volatiles less abundant than siderophiles. In contrast to the mare data in Fig. 2.10, the highland data have not been corrected for an indigenous contribution (from Anders *et al.*, 1973).

requires projectiles of at least decimeter size (Anders *et al.*, 1973). Consequently the ancient component shows much more clearly in separates of coarser materials hand picked from the soils (Figure 2.13). The most striking feature of the ancient component is the abundance of volatiles compared to siderophiles. The ancient meteoritic component thus, in contrast to the more recent micrometeoroid component, appears to be fractionated.

Two types of ancient components may be present and differ in the proportion of siderophile elements. In Figure 2.13 histograms of Ir/Au and Re/Au clearly show the two groups. Because the original data from Apollo 14 showed a tendency for light and dark norites to fall in the lower and upper groups respectively, the groups have been referred to as LN and DN (Morgan *et al.*, 1972). Both groups may consist of subgroups although data are limited. The LN component is the most abundant. The ancient components do not match any of the known chondritic meteoroid components. They are too low in Sb, Se, and Ag and in the case of the LN component they are too low in Ir and Re. The deficiencies can not be explained by volatilization either during impact or metamorphism. Similarly there is no obvious connection with iron meteorites.

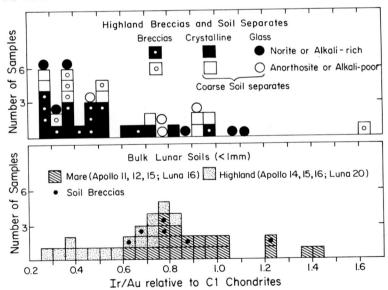

Fig. 2.13. (Top) At least two types of ancient meteoritic component seem to exist. The LN group, comprising the majority of samples, is strongly depleted in refractory siderophiles (Ir, Re) relative to Cl chondrites. The DN group on the right shows a lesser depletion. (Bottom) Most lunar soils, especially those from the highlands, have Ir/Au ratios below the Cl value (1.00 in this normalization). This reflects admixture of the ancient meteoritic component with its characteristically low Ir/Au value (from Anders *et al.*, 1973).

Anders *et al.*, (1973) conclude by suggesting that planetesimals have the following compositional traits.

1. They apparently contained nearly their cosmic complement of siderophiles. Thus they must have been independently formed bodies, not cast-offs of a larger body that had undergone a 'planetary' segregation of metal and silicate. Such segregation usually depletes siderophiles by factors of $10^{-3} - 10^{-4}$ (Anders *et al.*, 1971; Laul *et al.*, 1972b). Hence the planetesimals cannot represent material spun off the earth after core formation (Wise 1963, O'Keefe, 1969, 1970), condensates of volatilization residues from a hot Earth (Ringwood, 1966, 1970), or fragments of a differentiated proto-moon disrupted during capture (Urey and MacDonald, 1971). If their Fe content deviated at all from the cosmic abundances, this must be attributed to a 'nebular' metal-silicate fractionalization, by analogy with meteorites and planets (Urey, 1952; Wood, 1962; Larimer and Anders, 1967, 1970; Anders, 1971). This process, presumably based on the ferromagnetism of metal grains, has fractionated metal from silicate by factors of up to 3 in the inner solar system.

2. Volatile elements were depleted to <0.1 their cosmic abundance, as in the eucrites and the earth and moon as a whole. In terms of the two-component model of planet formation (Larimer and Anders, 1967, 1970), the planetesimals thus contained less than 10 per cent of low-temperature, volatile-rich material; the remainder consisted of high-temperature, volatile-poor material.

3. In the LN (but not the DN) planetesimals, refractory metals (Ir,Re) were depleted by 60-70 per cent. This raises the possibility that refractory oxides (CaO, Al_2O_3, etc.) were likewise depleted, because these elements tend to correlate in cosmo-chemical fractionation processes. No direct evidence on this point is available from the lunar samples, because lunar surface rocks themselves are enriched in refractory oxides. However, since the moon as a whole is strongly enriched in refractories (Gast, 1972), it would seen that only the DN, not the LN planetesimals can be important building blocks of the moon. The LN planetesimals are chemically complementary to the bulk of the moon; a fact of profound if obscure significance. It is most likely that planetesimals comprise two of more meteoritic types which are no longer represented among present-day meteorite falls.

History of the Meteoroid Flux

The marked difference in crater density between the mare and highland areas of the moon has long suggested that the flux of meteoroids at the lunar surface has changed with time. Urey (1952), Kuiper (1954), Kuiper *et al.* (1966) and Hartmann (1966) had all postulated an early intense bombardment of the moon. However, until the Apollo missions it was not possible to independently date the cratered surfaces and these arguments remained purely speculative. The dating of several lunar surfaces upheld the earlier hypotheses and it was found that the older mare surfaces had in fact caught the tail-end of the rapidly declining early intense meteoroid bombardment (Hartmann, 1970) (Fig. 2.14). The decline of the flux in the first 10^9 years was very rapid with a half-life of approximately 10^8 yr. The flux since 3.3 aeons has been relatively constant and may be increasing slightly.

The lunar meteoroid flux may have evolved in three phases (Hartmann, 1970):

1.   An intense bombardment of low velocity (1.7–2 km s^{-1}) circumterrestrial particles left over after the formation of the moon.

2.   An intense bombardment of the last planet-forming planetesimals swept up from low-eccentricity solar orbits with collision half-lives of the order of 10^8 yr. These produce medium velocity collisions (2–10 km s^{-1}) and probably resulted in the formation of the circular mare.

3.   The present cratering phase of high velocity (8–40 km s^{-1}) sporadic meteorites and cometary materials.

Phase 1 would necessarily be short-lived, since the moon moved under tidal influence to half its present distance in less than 10^8 years, and since many circumterrestrial particles would escape into circumsolar orbits. Most low-velocity particles "leading" out of the earth-moon system would probably be swept up in less than 10^6 years (Arnold, 1965). Some originally circumterrestrial particles could, however, be converted by perturbations into higher velocity particles of phase 2 (Arnold, 1965; Öpik, 1966).

The hypothesis of two phases (1 and 2) for the early bombardment might help explain why some ancient craters and basins suggest low-velocity impacts, while other more recent ones (Orientale) suggest high-velocity impacts; different impact velocities might also account for differences between craters and basins of equal size, some having only a single rim and some having multiple concentric ring systems.

The major crystalline breccia units such as the Fra Mauro Formation,

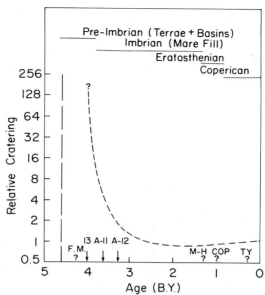

Fig. 2.14. Relative cratering rate (present day = 1) as a function of time. The curve is based on dated surfaces in Mare Tranquilitatis (A-11), Mare Procellarum (A-12), Fra Mauro (F.M.), Marius Hills (M-H), Copernicus (COP), and Tycho (Ty) (from Hinners, 1971).

probably formed as a consequence of the excavation of the circular mare during phase 2. That is, most of the large scale stratigraphy and topography of the lunar nearside formed during phase 2. Tera *et al.* (1974) view phase 2 as a cataclysm during which cratering rates increased sharply.

Phase 3 extends from approximately 2.0 or 2.5 aeons to present and accounts for the finishing touches to the surface morphology of the moon and is responsible for most of the development of the lunar soil particularly in mare areas. There might have been a slight increase in the flux during phase 3 as a result of asteroid collisions (Hartmann, 1970). More recently Hartung and Störzer (1974) have presented some evidence to suggest that the flux has been increasing over at least the last 10^4 yr. Their results are based upon solar flare track densities in impact produced glass lined pits. Their data indicate that the age distribution of impacts formed during the last 10^4 yr can be represented by an exponential curve corresponding to a doubling in the microcrater production rate about every 2000—3000 yr. (Figure 2.15). The present meteoroid flux appears to be in a quasiequilibrium such that it varies with time in response to the different processes producing and destroying small particles in the solar system. The present increase may be due to the arrival of Comet Encke which has been present in the solar system for several thousand years (Whipple, 1967).

More direct evidence of fluctuations in the meteoroid flux has come

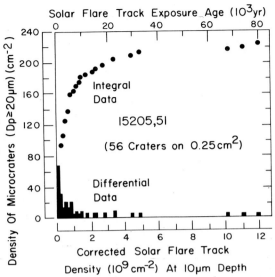

Fig. 2.15. Differential and integral age distributions for microcraters on sample 15205. The differential data indicate directly the *instantaneous* microcrater production rate (per 1000 yr) as a function of time in the past (exposure age). The slope of a line from the origin to any integral data point is the *average* microcrater production rate over the corresponding time (exposure age) for that data point. (from Hartung and Störzer, 1974, *Proc. 5th Lunar Sci. Conf., Suppl. 5, Vol. 3*, Pergamon).

from the distribution of complex impact-produced glass particles (agglutinates) in the lunar soil (Lindsay and Srnka, 1975). The slope of the log-number-flux vs log-mass plot of meteoroids in the mass range $10^{-5.7}$ to $10^{-7.1}$ g varied in a cyclical fashion with a depth in the lunar soil (time). The cycles may reflect an increase in the small particle flux due to the passage of the solar system through the spiral galactic arms in periods of the order of 10^8 yr. Phase 3 can thus be regarded as a period during which the meteoroid flux is, by comparison with phases 1 and 2, relatively small and in a cyclical quasiequilibrium state.

Meteoroid Energy

It is clear from the preceding section that the energy flux from meteoroids impacting the lunar surface has changed considerably with time. Even if we assume that the shape of the mass-frequency distribution has remained constant the energy flux was at least 2 orders of magnitude greater during much of the pre-mare time than at present. It is also apparent that during phases 1 and 2 of the flux history there were greater numbers of very-large impacts suggesting that the energy flux may have been even greater and also more effective in shaping the lunar surface than a model based on the present mass-frequency distribution shows.

Since phase 3 of the flux history covers most of post-mare time, which accounts for over 60 per cent of lunar history, and the time during which much of the lunar soil evolved, it is instructive to look at estimates of the current meteoroid energy flux. Iridium and osmium are rare in terrestrial sediments and comparatively more abundant in meteorites such that their concentrations can be used as an index to the amount of meteoritic material included in deep-sea sediment (Barker and Anders, 1968). By accepting the chemical composition of Cl carbonaceous chondrites as being representative of non-volatile cosmic matter Barker and Anders (1968) found the terrestrial accretion rate for meteoritic material to be 1.2 ($\pm$0.6) x 10^{-8} g cm^{-2} yr^{-1} over the past 10^5 to 10^6 yr. They were also able to set a stringent upper limit on the accretion rate by assuming that all Os and Ir in the sediments was cosmic. The upper limit is 2.9 x 10^{-8} g cm^{-2} yr^{-1}. The flux on the earth is enhanced by gravitational focusing (f = 1.8) by comparison with the moon (f = 1.03) but the difference is well within the uncertainties attached to the above estimates.

Similarly trace element enrichment in lunar soil and soil breccia samples from the Apollo 11 site indicated a mass flux of 3.8 x 10^{-9} gm c^{-2} yr^{-1} (Ganapathy *et al.*, 1970a). Subsequently Ganapathy *et al.* (1970a, b) estimated the mass flux at 4 x 10^{-9} g cm^{-2} yr^{-1} using similar trace element studies and soil samples from the Apollo 12 site. The Apollo 11 and 12 sites are both located on mare areas so that the flux estimates probably represent estimates of Hartmann's (1970) phase 3 of the flux history.

Dohnanyi (1971) used the estimates of the modern mass frequency distribution to arrive at an integrated mass estimate of 2 x 10^{-9} cm^{-2} yr^{-1} for the moon. Gault *et al.* (1972) arrived at a similar figure and later refined the figure to 1.4 x 10^{-9} g cm^{-2} yr^{-1} (Gault *et al.*, 1974).

From the available information and assuming a root-mean-square velocity of 20 km s^{-1} the post-mare energy flux from impacting meteoroids at the lunar surface has been between 2 to 8 x 10^3 ergs cm^{-2} yr^{-1}. By comparison with the terrestrial environment this energy flux is extremely small. However, the lack of an atmosphere and the large amount of geologic time available make meteoroid impact the most effective sedimentary process on the lunar surface. The partitioning of the flux energy is complex and is dealt with in some detail in Chapter 3.

Solar Energy

At the avarage distance of the earth-moon system from the sun the radiant energy flux is 4.4×10^{13} ergs cm^{-2} yr^{-1} at a normal angle of incidence. On the earth this energy fuels a global heat engine which provides most of the chemical and mechanical energy for producing and transporting sedimentary materials. Re-direction of radiant energy into sedimentary processes on the earth is dependent upon fluid intermediaries — the atmosphere and hydrosphere. Without fluid intermediaries solar energy is almost totally ineffective — such is the case on the moon.

The radiant-energy flux contributes an average of 1.1×10^{13} ergs cm^{-2} yr^{-1} to the lunar surface. That is, at least 10 orders of magnitude larger than the meteoroid flux contribution. Most of this energy is simply reradiated into space during the lunar nights. However, a small, possibly significant, amount of solar energy is effective in the sedimentary environment.

Electrostatic Transport

Electrostatic transport of lunar surface materials was first suggested by Gold (1955). The proposed driving mechanism was transient charge separation due to photoelectron ejection from fully sunlit surfaces. However, later theoretical studies suggested that under daytime conditions soil grains would rise only a few millimeters at most above the surface (Gold, 1962, 1966; Singer and Walker, 1962; Heffner, 1965). Subsequently, Surveyor spacecraft recorded a bright glow along the western lunar horizon for some time after sunset suggesting that a tenuous cloud of dust particles extends for up to 30 cm above the lunar surface during and after lunar sunset. Criswell (1972) and Rennilson and Criswell (1974) suggest that this was due to the retention of large differences of electrical surface charge across light/dark boundaries due to the highly resistive nature of the lunar surface. The dust motion could result in an annual churing rate of 10^{-3} g cm^{-2} at the lunar surface which is about 4 orders of magnitude greater than the rate at which soil accumulates from meteoroid impact erosion (Criswell, 1972). The horizon glow observed by the Surveyor spacecraft appears to be produced by particles with a diameter of 10μm. As discussed in Chapter 6, such particles form about 10 per cent of the soil mass. The concept of electrostatic transport of lunar materials thus appears to be very model-dependent, but is potentially capable of transporting very large volumes of lunar soil.

Thermal Effects

Thermal extremes in the terrestrial environment, particularly in hot or cold deserts, are important in shattering rock exposures and producing sedimentary materials. However, despite the dryness of these terrestrial environments, it appears that expanding and contracting water, even in small amounts, is the critical agent in the shattering of terrestrial rocks. McDonnell *et al.* (1972, 1974) found no evidence of any degradation of lunar rocks cycled from -186°C to +100°C in vacuum. The conditions of their experiments were thermally equivalent to the lunar day-night cycle which led them to conclude that thermal cycling is at best a very weak erosional mechanism. In contrast, thermally triggered slumping of lunar soil appears to be a relatively common occurence on the lunar surface (Duennebier and Sutton, 1974). Such small scale slumping may be an important erosional mechanism. However, gravity is the main source of energy for such mass movements; it is discussed in a following section. Duennebier and Sutton (1974) have also suggested that thermal moonquakes generated by rock fracturing could cause a gradual decrease in grain size by grinding on slip boundaries. However, they were unable to evaluate the importance of the mechanism.

Solar Wind Sputtering

The small but persistent solar-wind flux on the lunar surface causes small amounts of erosion. Hydrogen, with a flux of 2.0×10^8 ions cm^{-2} s^{-1}, produces about 70 per cent of the sputtering erosion and helium, with a flux of 9.0×10^6 ions cm^{-2} s^{-1}, the remaining 30 per cent (McDonnell and Ashworth, 1972). Experimental studies by McDonnell *et al.* (1972) and McDonnell and Flavill (1974) simulating solar wind with a velocity of 400 km s^{-1} show that there is a perferential loss of rock material on the rims of microcraters and an etching of mineral grain boundaries and spall zone faults caused by hypervelocity impacts. This results in an enhancement of these features. One of the striking features of solar wind sputtering is the formation of discrete needles aligned in the direction of the particle source (Fig. 2.16). McDonnell and Flavill (1974) view these features as being the result of "topology instability" where a flat surface with small irregularities develops quasistable shapes of high inclination, which propagate themselves under sputtering.

Computer simulation studies suggests that the observed topologic effect of solar wind sputtering results mostly from the relatively high angular dependence of the sputter yield (Fig. 2.17). That is, inclined surfaces recede faster than those normal to the source direction. On the lunar surface needles are therefore only likely to develop where the incidence angle of the solar wind is restricted to a relatively small range such as in the base of a deep void.

The erosion caused by solar wind sputtering is small; probably of the

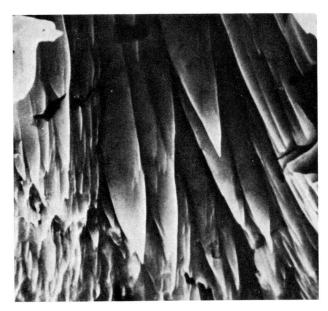

Fig. 2.16. Enlargement of a section of helium sputtered crystalline breccia showing the development of needles aligned towards the incident sputtering beam (Magnification 2000) (from McDonnell and Flavill, 1974, *Proc. 5th Lunar Sci. Conf, Suppl. 5, Vol. 3*, Pergamon).

order of $0.043 \pm 0.010\,\text{Å}\,\text{yr}^{-1}$ for lunar crystalling breccia (McDonnell and Flavill, 1974). At an average rock density of $3\ \text{g cm}^{-3}$ this corresponds to a loss of $1.3 \times 10^{-9}\ \text{g cm}^{-2}\,\text{yr}^{-1}$ which is four orders of magnitude smaller than the total flux of materials produced by the current meteoroid flux. The effects of solar wind sputtering are thus not of great importance in terms of shaping lunar landscape and the stratigraphic record. However, sputtering has a significant effect in destroying the record of micrometeoroid impacts and high energy particle tracks. This may have resulted in an underestimate of the influx of micrometeoroids smaller than 10^{-6} g (Fig. 2.18).

Gravitational Energy and Mass Wasting

Predictably, evidence of mass movement is abundant on the moon. Large scale slump structures are visible on the inner walls of most large craters and in some cases large boulder streams are visible even from lunar orbit. Similarly on steeper slopes large boulders may be seen lying at the downslope ends of long trails gouged into the lunar soil (Fig. 2.19) and in many highland areas "patterned ground" consisting of linear features trending down slope is readily visible.

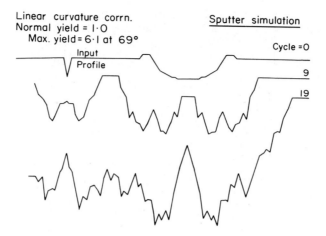

Fig. 2.17. Computer simulation of the sputter erosion of a small crater and a V-shaped irregularity. The instability of this topography, when an appropriate angular dependence of the sputter efficiency is incorporated, demonstrates the very close approach to the needle profiles observed in actual sputtering (from McDonnell and Flavill, 1974, *Proc. 5th Lunar Sci. Conf., Suppl. 5, Vol 3*, Pergamon).

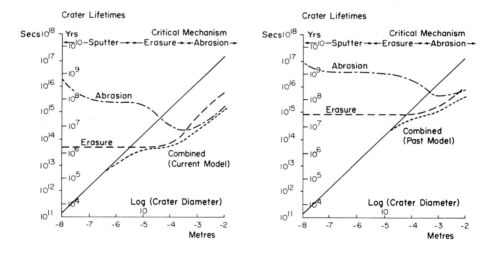

Fig. 2.18. Crater lifetimes for an exposed rock on the lunar surface due to solar wind sputtering, impact abrasion by craters smaller than the crater diameter and impact erasure by craters larger than the crater diameter. A combined lifetime is also shown. The left figure is computed from the current flux at 1 AU heliocentric distance and the right figure from the "past" flux which is required to establish observed lunar equilibrium surfaces. This is deduced without recourse to rock dating techniques (from McDonnell and Flavill, 1974, *Proc. 5th Lunar Sci. Conf., Suppl 5, Vol. 3*, Pergamon).

Fig. 2.19. Boulder trails on steep highland slopes at the Apollo 17 landing site (NASA Photo AS17-144-21991).

Mass Wasting of Soil

Considerably more detailed and subtle information concerning mass wasting was obtained during the Apollo missions. At the Apollo 15 site it was found that on steep slopes of the Apennine Front boulders were scarce on the lower slopes of the mountains but the upper slopes were often covered by boulders or consisted of blocky exposures (Fig. 2.20) (Swann *et al.*, 1972). This implies that the soil is preferentially thinned between the mountain top and the steep face. This zone probably occurs because ejecta from craters on the flat hill top are distributed randomly around the source craters, whereas ejecta on the slope are distributed preferentially down slope. Hence, a zone of disequilibrium exists high on the slope where material is lost down slope more rapidly than it is replenished by impacts. The effects of this downslope movement are shown clearly in Figure 2.21 where dark Imbrian ejecta can be seen streaming down the lighter slopes of the Apennine Front.

A similar pattern suggesting mass wasting is very clear at Hadley Rille (Swann *et al.*, 1972). As the rille is approached from the crest of the rim, the surface slopes gently downwards and the soil thins and becomes coarser.

Fig. 2.20. The blocky upper slopes of Mount Hadley at the Apollo 15 landing site. Mass wasting has resulted in thinner soil on the higher steeper mountain slopes (NASA Photo AS15-84-11304).

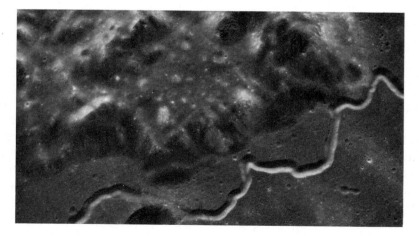

Fig. 2.21. The Apennine Front near the Apollo 15 landing site showing the streaming of dark colored materials down slope by mass movement (NASA Photo AS15-9814).

Similarly the composition of the soil changes and becomes richer in mineral grains and deficient in glass fragments (Lindsay, 1972). Within about 25 m of the lip of the rille, soil is essentially absent so that numerous boulders and bedrock exposures are to be seen (Fig. 2.22). This thinning is inferred to be the result of near-rim impacts that distributed material in all directions including into the rille. However, the narrow zone of thin soil along the rille receives material only from the east because impacts that occur within the rille to the west do not eject material up to the rim (Fig. 2.23). This ejection pattern results in the loss of material toward the direction of the rille; therefore, as the rille rim recedes backwards by erosion, so will the zone of thin regolith recede.

·Rock fragments are more abundant in the vicinity of the Hadley Rille rim than they are on the mare surface to the east. The increase becomes noticeable approximately 200 to 300 m east of the lip of the rille. Most of the fragments at a distance of 200 to 300 m east are a few centimeters across. The size of the fragments increases markedly as the surface begins to slope gently down toward the rille; bedrock is reached at the lip.

The abundance of rocks in the 200- to 300- m zone along the rim of Hadley Rille is related to the nearness to the surface of outcrops in the

Fig. 2.22. Large blocks and exposures of bedrock at the lip of Hadley Rille (NASA Photo AS15-82-11147).

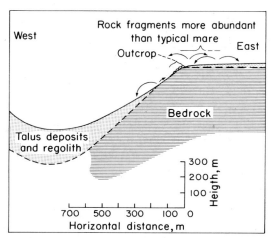

Fig. 2.23. Diagram to illustrate winnowing of lunar soil into a rille. Impacts on the rille rim eject material in all directions but the rim receives ejecta from only one side. This results in a net movement of soil into the rille (after Swann *et al.*, 1972).

vicinity of the rim. In the narrow zone along the lip all craters greater than 0.5 m or so in diameter penetrate the fine-grained material, and therefore the ejecta consist primarily of rock fragments. In the areas of normal soil thickness, only those craters greater that 20 to 25 m in diameter penetrate the regolith, and even then most of the ejecta are fine-grained materials from craters approximately 100 m in diameter. Therefore, the blocky nature of the 200 to 300 m zone along the rille is due to the nearby source of rocks in the area of very thin soil along the rille rim.

At the bottom of Hadley Rille, the fragment-size has a bimodal distribution (Fig. 2.24). Numerous large boulders have broken from the outcrops at the rille rim and were large enough to roll over the fines to the bottom. The rest of the material is mainly fine grained and has probably been winnowed into the rille by cratering processes. Where the rille meets the pre-mare massif, the rille wall is made of fine-grained debris, and the bottom of the rille is shallower and flatter than elsewhere. This situation indicates a considerable fill of fine-grained debris derived from the massif. Mass movement is thus filling the rille and rounding the v-shape.

Talus Slopes

The Hadley Rille at the Apollo 15 landing site also offers the best information about talus slopes on the moon. The talus slopes that form the main walls of the rille are blocky compared to most of the lunar surface (Fig. 2.24). Loose debris is approximately at the angle of repose. Boulder tracks visible on the slope of the rille wall indicate recent instability of the blocks.

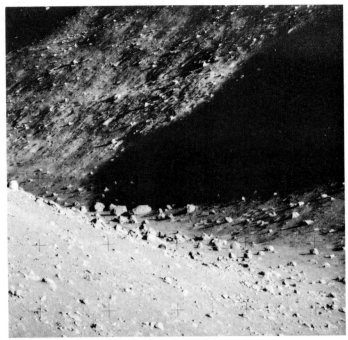

Fig. 2.24. Soil and boulders accumulated in the floor of Hadley Rille. Note the extreme bimodality of the particle size distribution as a consequence of mass wasting (NASA Photo AS15-84-11287).

The talus is especially blocky where penetrated by fresh craters. The obvious source of most talus is the outcropping ledges of mare basalt near the top of the wall. Loose blocks that lie above the level of outcrops can be accounted for as blocks produced by impact reworking of the lunar surface.

The blocky talus deposits are generally poorly sorted due to a larger component of fine-grained debris as compared to talus slopes on earth. This difference is undoubtably caused by impact comminution of the lunar talus and to the addition of fine-grained ejecta from the mare surface beyond the outcrops. Many patches of talus in the rille are so recently accumulated, however, that fine-grained debris does not fill the interstices (Fig. 2.25). Where aligned with fractured outcrops, some of these block fields appear jumbled but not moved far from their source, similar to fields of frost-heaved blocks that cover outcrops on some terrestrial mountain peaks. On the northeast wall of Hadley Rille, blocks commonly accumulate on a bench just below the outcrop scarp; farther down the wall, blocks are in places concentrated on the steep lower part of convexities in the slope. A few patches of blocks are elongate down the slope similar to stone stripes on earth. Near the top of the rille wall, horizontal lines of blocks underlie finer regolith in places and represent rocks that are apparently close to their

Fig. 2.25. A blocky cratered area on the lip of Hadley Rille (NASA Photo AS15-82-11082).

bedrock source.

Some talus blocks at Hadley Rille are more than 10 m across (Fig. 2.24). The largest blocks are about the same size as, or a little thicker than, the largest unbroken outcrops. Unbroken blocks of basalt this large are uncommon on earth, demonstrating that these lunar basaltic flows are both thick and remarkably unjointed compared to many terrestrial counterparts (Swann *et al.*, 1972).

The talus blocks have a range of shapes and textures which may provide an understanding of lunar weathering and erosional processes. Many large talus blocks have split in their present location. Other evidence of lunar weathering and erosional processes on the blocks include expressions of layering and rounding of many blocks.

The thickness of the talus deposits is not known, nor is the distance that the lip of the rille has receded. The present profile appears to be a consequence of wall recession by mass wasting so that the talus aprons of the two sides coalesce.

Soil Creep

Direct evidence of mass wasting is available from seismic records. Many thousands of seismic events have been recorded by the short-period components of the Apollo passive seismic network. The majority of these events appear to be small local moonquakes triggered by diurnal temperature changes. These events, which have been called thermal moonquakes, are recognized by the repetition of nearly identical signals at the same time each lunation (Fig. 2.26) (Duennebier and Sutton, 1974). Thermal moonquake activity begins abruptly about 2 days after the lunar sunrise and decreases rapidly after sunset. Thermal moonquakes appear to have source energies of between 7 x 10^7 and 10^{10} ergs within a 1 to 4 km range.

There is considerable uncertainty attached to these estimates of source energies. Erosion rates and source energies estimates depend largely on the coupling efficiency which has been estimated from explosions set off on the lunar surface for the active seismic experiments (Kovach et al., 1972). Criswell and Lindsay (1974) have argued that the erosion rates implied by these estimates of the coupling factor are much too large by several orders of magnitude. Their argument is based on a rare terrestrial phenomenom known as booming sands (Lindsay, et al., 1975; Criswell et al., 1975). Booming

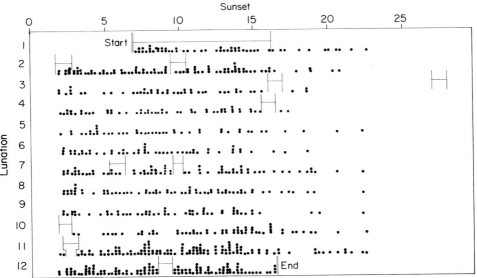

Fig. 2.26. Occurence calendar of all thermal moonquakes observed at station 14. This figure is read like a calendar; each line represents one lunation. Each lunation is broken into terrestrial days, starting at sunrise. Dots represent the number of thermal moonquakes occuring in each 6-hour period. Note that activity drops off after sunset and that no thermal moonquakes are observed during the latter part of the lunation. Solid horizontal lines show periods when the event detection program was not run. Dotted horizontal lines show periods when analog records are available. Sensitivity of detection program was higher during lunations 11 and 12. The recording period starts on February 11, 1971, and ends on January 10, 1972 (from Duennebier and Sutton, 1974).

sands emit a low frequency (f<100 Hz) sound during slumping or avalanching. Very efficient (1 per cent) conversion of slumping energy to seismic energy has been observed (Lindsay *et al.*, 1975; Criswell *et al.*, 1975). Booming efficiently produces seismic signals in the 50 to 80 Hz and less efficiency in the 10 Hz range. Thermal moonquakes have a narrow frequency distribution much as the booming sands. Some thermal moonquakes occur as an evolving sequence during the lunar day that is very faithfully repeated from one day to the next. In the booming process local slopes would control the spectral outputs of each event and the very slight displacements required for each event to occur would permit the slumping of each element to continue over many cycles before significant local slope changes occured. Sound production of booming sands appears to relate to the mechanical coupling between grains. Consequently, in the terrestrial environment, booming only occurs in sands that are extremely dry and where the surfaces of the grains are very smooth. The extreme dryness of the moon precludes the need for preconditioning of the grains and as a consequence booming, which appears to be rare on the earth, may be very common in the lunar vacuum. However, considerably more data are required before the mechanism is well enough understood to apply it to the lunar case.

If Duennnebier and Sutton (1970) are correct and we assume an energy release of 5×10^8 ergs for an average thermal moonquake, a radius of detectability of 5 km, and about 50 thermal moonquakes per lunation, then 10^9 ergs km^{-2} are released each lunation. This corresponds to an average of 2.63×10^{-2} erg cm^{-2} yr^{-1}.

Clearly a significant amount of potential energy is released each year at the lunar surface and it must play an important role in the evolution of the lunar surface. The question is how? Obviously mass wasting is important in transporting material down slope towards some imaginary base level — perhaps the mean selenoid. Possibly much more important, however, is the role mass wasting plays in conjunction with the meteoroid flux. The rate at which mass wasting proceeds is a direct function of slope. That is, the steeper the slope the more rapidly mass wasting removes soil from its surface. The lunar soil appears to be produced largely by meteoroid impact. However, soil production is a self-damping process. As the thickness of the soil blanket increases the ability of meteoroids to penetrate to bedrock is reduced. That is, larger and larger meteoroids are necessary to penetrate the soil and excavate new bedrock materials. Consequently, more and more kinetic energy is expended reworking the soil and less and less is expended on erosion. On steep slopes the effects of blanketing will be, in part, removed as the soil migrates down slope leaving a thinner soil blanket to impede erosion. Thus, meteoroid erosion will cause surfaces to migrate at a rate proportional to this slope. Ultimately this produces a more subdued topography as the upper slopes of mountain fronts are eroded while the lower slopes are buried. The accumulation of soil blankets is treated in more detail in Chapter 6.

Internal Energy (The Volcanic Contribution)

The flooding of the lunar mare by basaltic lava flows was one of the most important late stage lunar events in terms of both the evolution of the moon and the development of the stratigraphic record. The stratigraphic relationships of the mare basalts are discussed in more detail in Chapter 4. The contribution of pyroclastic materials to the lunar sedimentary record is, however, much more difficult to evaluate. Numerous craters and crater chains on the moon have at some time been attributed to volcanism (for example, Green, 1971). In most cases it is difficult or impossible to substantiate such claims. However, some large conical hills with summit depressions such as those in the floor of the crater Copernicus are difficult to explain in any other way except as internal in origin (Fig. 2.27).

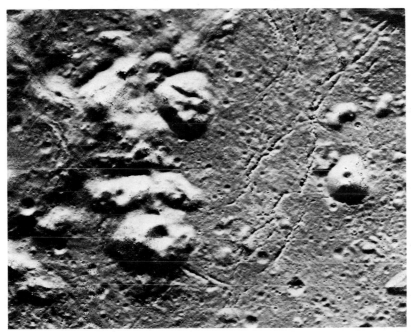

Fig. 2.27. Conical hills with summit depressions which suggest an internal or volcanic origin. These features occur in the floor of crater Copernicus (NASA Photo Lunar Orbiter V-154-H₁). Framelets are 440 m wide.

In most of the mare, particularly around the edges, there are areas of low albedo which are generally referred to as "dark mantle material". In some of these areas small craters with dark halos can be seen. It had been suggested that these craters were cinder cones and the dark mantle was a blanket of dark pyroclastic material presumably relating to the very youngest stage of mare or post-mare volcanism. One of these areas, the Taurus-Littrow Valley on the southeastern edge of Mare Serenitatis, was selected as the Apollo 17 landing site with the express purpose of studying the dark mantle. As with many albedo changes observed from lunar orbit the dark mantle proved difficult to identify on the ground. The dark halo crater (Shorty Crater) visited at the Apollo 17 site proved to be an impact crater and would have been relatively unimportant except for the discovery of a rusty-orange soil along its rim.

The significance of the orange soil from the Apollo 17 site, and similar emerald green-glasses from the Apollo 15 site, in relation to dark mantle materials is discussed in detail in Chapter 4. Briefly however, the consensus among research groups at the present time is that both the orange and green glasses are pyroclastic in origin. The glasses have compositions and ages similar to the associated mare basalts and have been interpreted as the product of lava fountaining (Reid et al., 1973; McKay and Heiken, 1973; Carter et al., 1973). Green glass forms up to 20 per cent of some Apollo 15 soils and in general increases in abundance towards the Apennine Front (Reid et al., 1973) (Fig. 2.28). The orange glass forms from 6 to 26 per cent of typical soils from the Apollo 17 site (Rhodes et al., 1974). Overall the contribution of pyroclastic material does not appear to be large but locally it may form a significant proportion of the soil, particularly around the edge of the mare. Because the pyroclastic materials are of the same age as the basaltic substrate they may have been exposed to reworking in the lunar soil for much of the soil's history. It is thus possib'e that pyroclastic materials are much more abundant in the soil than we realize and are simply modified beyond recognition.

Other Energy Sources

During several of the lunar missions and particularly during Apollo 17 the astronauts observed bright streamers accompanying spacecraft sunrise (Fig. 2.29). The streamers are similar to those observed at sunset on earth. Initially the streamers were interpreted as being solar in origin (Bohlin, 1971) but more recently arguments have been presented which suggest that they were produced by light scattering in the vicinity of the moon (McCoy

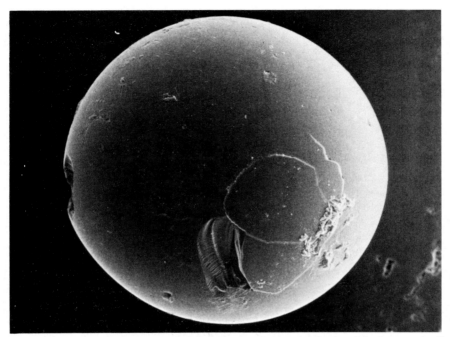

Fig. 2.28. A green glass sphere (390 μm diameter) of probable pyroclastic origin from an Apollo 15 soil sample (NASA Photo S-72-53599).

and Criswell, 1974). The most reasonable model for the light scattering involves particles with a diameter of 0.1 μm and a number density of 10^{-1} cm^{-3} at 1 km altitude, 10^{-2} cm^{-3} at 10 km altitude and between 10^{-5} to 10^{-6} cm^{-3} at 100-200 km altitude (Fig. 2.30). Number densities of these magnitudes are in excess of those predicted for secondary meteoritic particles (Gault et al., 1963). The number density of particles is apparently variable over time periods of the order of 6 months as the sunrise streamers were not seen by all Apollo missions. Local variations in number density are also suggested by rapid changes in the intensity of the streamers during transit.

McCoy and Criswell (1974) found no evidence to suggest an origin for the particles. However, if the particles are of lunar origin and are in suborbital ballistic trajectories (70 s flight time) the maximum churning rate would be 9.5 x 10^{-5} g cm^{-2} yr^{-1}. Such small particles could be swept away from the moon by radiation pressure resulting in a net loss from the moon. Some of the particles may be in lunar orbit in which case the churning rate would be much reduced. The evidence that McCoy and Criswell (1974) present to suggest dust at high altitude is compelling. However, the effectiveness of whatever the churning mechanism might be is very model dependent. If these estimates are correct there may be a mechanism available

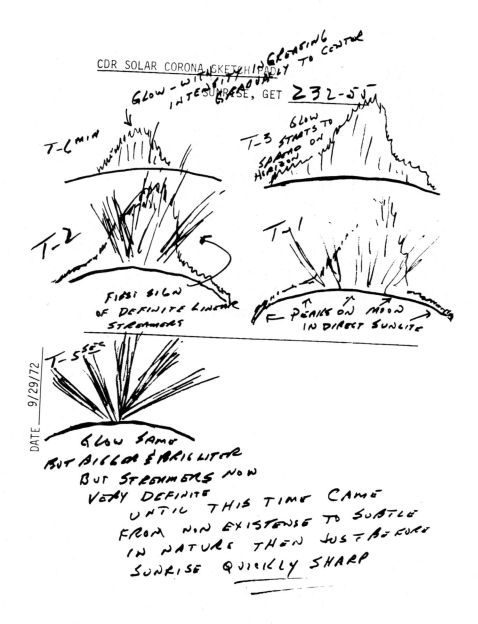

Fig. 2.29. Five sketches drawn by E. A. Cernan of sunrise as viewed from lunar orbit during the Apollo 17 mission. The times in minutes (i.e. T-6, T-3, T-2, and T-1 min) and seconds (i.e. T-5 sec) refer to the time before first appearance of the sun (NASA Photo S-73-15138, from McCoy and Criswell, 1974).

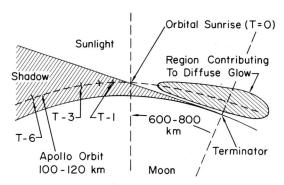

Fig. 2.30. A schematic cross section (approximately along a lunar longitude) of the moon in the plane of the spacecraft orbit (dashed line). Spacecraft proceeds from left to right approaching the terminator at 3° min-1 or approximately 1.6 km sec-1 (from McCoy and Criswell, 1974, *Proc. 5th Lunar Sci. Conf., Suppl. 5, Vol. 3*, Pergamon).

on the lunar surface to transport fine particles at a rate of two orders of magnitude faster than meteoroid impact produces them.

Concluding Remarks

It is clear from the preceding discussion that the critical difference between terrestrial and lunar sedimentary processes results from the lack of fluids on the lunar surface. Without fluid intermediaries solar energy, the dominant energy source in the terrestrial environment, becomes largely ineffective as a source of erosive energy in the lunar environment. Meteoritic energy, a feeble energy source in the terrestrial environment, thus becomes by default the main source of erosional and transportational energy. The importance of gravitational energy as an agent of transport is probably small, but, by moving loose material from steep slopes it improves the erosional efficiency of smaller meteoroid impacts. The only other obvious source of detrital materials is pyroclastic. Locally pyroclastic materials are important but it is difficult to evaluate their overall significance on the lunar surface as they have probably undergone considerably reworking which may have rendered much of the pyroclastic material unrecognizable.

Electrostatic transport and an unknown energy source which moves dust to orbital altitudes may be important transport mechanisms when particles smaller than 10 μm are considered. However, ultimately most of the clastic materials on the lunar surface are eroded and transported by meteoroid impact. Consequently, the following chapter is devoted entirely to an analysis of the ways in which meteoroid energy is partitioned to produce and transport detrital materials.

References

Allen, C. W., 1946. The spectrum of the corona at the eclipse of 1940 October 1. *Monthly Notices Roy. Astron. Soc., 106:* 137-150.

Anders, E., 1971. Meteorites and the early solar system. *Ann. Rev. Astron. Astrophys., 9:* 1-34.

Anders, E., Ganapathy, R., Keays, R. R., Laul, J. C. and Morgan, J. W., 1971. Volatile and siderophile elements in lunar rocks: Comparison with terrestrial and meteoritic basalts. *Proc. Second Lunar Sci. Conf., Suppl. 2, Geochim. Cosmochim. Acta, 2:* 1021-1036.

Anders, E., Ganapathy, R., Krahenbuhl, U. and Morgan, J. W., 1973. Meteoritic material on the moon. *The Moon, 8:* 3-24.

Arnold, J. R., 1965. The origin of meteorites as small bodies. II. *Astrophys. Jour., 141:* 1536-1547.

Bandermann, L. W., 1969. Interplanetary dust. In: H. Ogelman and J. R. Wayland (Editors), *High Energy Astrophysics*, NASA SP-199, 137-165.

Bandermann, L. W. and Singer, S. F., 1969. Interplanetary dust measurements near the earth. *Rev. Geophys., 7:* 759-797.

Barker, J. L., and Anders, E., 1968. Accretion rate of cosmic matter from iridium and osmium content of deep-sea sediments. *Geochim. Cosmochim. Acta, 32:* 627-645.

Berg, O. E. and Gerloff, U., 1970. Orbital elements of micrometeorites derived from Pioneer 8 measurements. *Jour. Geophys. Res., 75:* 6932-6939.

Berg, O. E. and Richardson, F. F., 1971. New and supplementary data from the pioneer cosmic dust experiments (abstract). *14th Plenary Meeting of COSPAR*, Seattle, Washington.

Bloch, M. R., Fechtig, H., Gentner, W., Neukum, G. and Schneider, E., 1971. Meteorite impact craters, crater simulations, and the meteoroid flux in the early solar system. *Proc. Second Lunar Sci. Conf., Suppl. 2, Geochim. Cosmochim. Acta, 3:* 2639-2652.

Bohlin, J. D., 1971. Photometry of the outer solar corona from lunar-based observations. *Solar Physics, 18:* 450-457.

Brown, H., 1960. Density and mass distribution of meteorites. *Jour. Geophys. Res., 75:* 1679-1683.

Carter, J. L., Taylor, H. C. and Padovani, E., 1973. Morphology and chemistry of particles from Apollo 17 soils 74220, 74241, 75081. *EOS, 54:* 582-584.

Chao, E. C. T., Borman, J. A., Minkin, J. A., James, O. B. and Desborough, G. A., 1970. Lunar glasses of impact origin: Physical and chemical characteristics and geologic implications. *Jour. Geophys. Res., 75:* 7445-7479.

Cour-Palais, B. G., 1969. Meteoroid environment model—1969 (near earth to lunar surface). NASA SP-8013, 30 pp.

Cour-Palais, B. G., Zook, H. A. and Flaherty, R. E., 1971. Meteoroid activity on the lunar surface from the Surveyor III sample examination. *14th Plenary Meeting of COSPAR*, Seattle, Washington.

Criswell, D. R., 1972. Lunar dust motion. *Proc. Third Lunar Sci. Conf., Suppl. 3, Geochim. Cosmochim. Acta, 3:* 2671-2680.

Criswell, D. R., Lindsay, J. F. and Reasoner, D. L., 1975. Seismic and acoustic emissions of a booming dune. *Jour. Geophys. Res., 80:* 4963-4974.

Dalton, C. D., 1969. Determination of meteoroid environments from photographic meteor data. NASA TR R-322, 134 pp.

Dohnanyi, J. W., 1970. On the origin and distribution of meteoroids. *Jour. Geophys. Res., 75:* 3468-3493.

Dohnanyi, J. W., 1971. Flux of micrometeoroids: Lunar sample analyses compared with flux models. *Science, 173:* 558.

Dohnanyi, J. W., 1972. Interplanetary objects in review: statistics of their masses and dynamics. *Icarus, 17:* 1-48.

Duennebier, F. and Sutton, G. H., 1974. Thermal moonquakes. *Jour. Geophys. Res., 79:* 4351-4363.

Erickson, J. E., 1968. Velocity distribution of sporadic photographic meteors. *Jour. Geophys. Res., 73:* 3721-3726.

Farlow, N. H. and Ferry, G. W., 1971. Cosmic dust in the mesophere. *14th Plenary Meeting of COSPAR*, Seattle, Washington.

Fechtig, H., 1971. Cosmic dust in the atmosphere and in the interplanetary space at 1 A.U. today and in the early solar system. In: C. L. Hemenway, P. M. Milman and A. F. Cook (Editors), *Evolutionary and Physical Problems of Meteoroids*. NASA SP-319, 209-222.

Ganapathy, R., Keays, R. R. and Anders, E., 1970a. Apollo 12 lunar samples: Trace element analysis of a core and the uniformity of the regolith. *Science, 170*: 533-535.

Ganapathy, R., Keays, R. R., Laul, J. C. and Anders, E., 1970b. Trace elements in Apollo 11 lunar rocks: Implications for meteorite flux and origin of moon. *Proc. Apollo 11 Lunar Sci. Conf., Suppl. 1, Geochim. Cosmochim. Acta, 1*: 1117-1142.

Gast, P. W., 1972. The chemical composition and structure of the moon. *The Moon, 5*: 121-148.

Gault, D. E., 1970. Saturation and equilibrium conditions for impact cratering on the lunar surface: Criteria and implications. *Radio Sci., 5*: 273-291.

Gault, D. E., Hörz, F. and Hartung, J., 1972. Effects of microcratering on the lunar surface. *Proc. Third Lunar Sci. Conf., Suppl. 3, Geochim. Cosmochim. Acta, 3*: 2713-2734.

Gault, D. E., Hörz, F., Brownlee, D. E. and Hartung, J. B., 1974. Mixing of the lunar regolith. *Proc. Fifth Lunar Sci. Conf., Suppl. 5, Geochim. Cosmochim. Acta, 3*: 2365-2386.

Gold, T., 1955. The lunar surface. *Monthly Nat. Roy. Astron. Soc., 115*: 585-604.

Gold, T., 1962. Processes on the Lunar Surface. In: Z. Kopal and Z. Michailov (Editors), *Proc. I.A.U., Pulkovo Meeting.*

Gold, T., 1966. The Moon's Surface. In: W. H. Hess, D. H. Menzel and J. A. O'Keefe (Editors), *Proc. I.A.U. Pulkovo Meeting.*

Green, J., 1971. Copernicus as a lunar caldera. *Jour. Geophys. Res., 76*: 5719-5731.

Hanappe, F., Vosters, M., Picciotto, E. and Deutsch, S., 1968. Chimie des neiges Antarctiques et faux de deposition de matiere extraterrestre-deuxieme article. *Earth Planet. Sci. Lett., 4*: 487-496.

Hartmann, W. K., 1965. Terrestrial and lunar flux of large meteorites in the last two billion years. *Icarus, 4*: 157-165.

Hartmann, W. K., 1966. Early lunar cratering. *Icarus, 5*: 406-418.

Hartmann, W. K., 1970. Preliminary note on lunar cratering rates and absolute time-scales. *Icarus, 12*: 131-133.

Hartung, J. B., Hörz, F. and Gault, D. E., 1971. Lunar rocks as meteoroid detectors. In: C. L. Hemenway, P. M. Milman and A. F. Cook (Editors), *Evolutionary and Physical Problems of Meteoroids*. NASA SP-319, 227-237.

Hartung, J. B., Hörz, F. and Gault, D. E., 1972. Lunar microcraters and interplanetary dust. *Proc. Third Lunar Sci. Conf., Suppl. 3, Geochim. Cosmochim. Acta, 3*: 2735-2753.

Hartung, J. B. and Störzer, D., 1974. Lunar microcraters and their solar flare track record. *Proc. Fifth Lunar Sci. Conf., Suppl. 5, Geochim. Cosmochim. Acta, 3*: 2527-2541.

Hawkins, G. S., 1963. Impacts on the earth and moon. *Nature, 197*: 781.

Hawkins, G. S. and Southworth, R. B., 1961. Orbital elements of meteors. *Smithson. Contrib. Astrophys., 4*: 85-95.

Heffner, 1965. Levitation of dust on the surface of the moon. *N. 66-16171, Minn. Univ. Rept.* TYCHO Meeting.

Hinners, N. W., 1971. The new moon; a view. *Rev. Geophys. Space Phys., 9*: 447-522.

Hörz, F., Brownlee, D. E., Fechtig, H., Hartung, J. B., Morrison, D. A., Neukum, G., Schneider, E., Vedder, J. F. and Gault, D. E., 1975. Lunar microcraters, implications for the micrometeoroid complex. *Planet. Space Sci., 23*: 151-172.

Hörz, F., Hartung, J. B., and Gault, D. E., 1971. Micrometeorite craters on lunar rock surfaces. *Jour. Geophys. Res., 76*: 5770-5798.

Humes, D. H., Alvarez, J. M., Kinard, W. H. and O'Neal, R. L., 1975. Pioneer 11 meteoroid detection experiment: Preliminary results. *Science, 188*: 473-474.

Jacchia, L. G., 1963. Meteors, meteorites and comets: Interrelationships. In: B. M. Middlehurst and G. P. Kuipers (Editors), *Moon, Meteorites and Comets*, 774-798.

Jacchia, L. G., and Whipple, F. L., 1961. Precision orbits of 413 photographic meteors. *Smithson. Contrib. Astrophys., 4*: 97-129.

Jennison, R. C., McDonnell, J. A. M. and Rogers, I., 1967. The Ariel II micrometeorite penetration measurements. *Proc. Roy. Soc. Lon., A., 300*: 251-269.

Keays, R. R., Ganapathy, R., Laul, J. C., Anders, E., Herzog, G. F. and Jeffery, D. M., 1970. Trace elements and radioactivity in lunar rocks: Implications for meteorite infall solar wind flux and formation conditions of moon. *Science, 167*: 490-493.

Kerridge, J. F., 1970. Micrometeorite environment at the earth's orbit. *Nature, 228*: 616-619.

Kerridge, J. F. and Vedder, J. F., 1972. Accretionary processes in the early solar system: An experimental approach. *Science, 177*: 161-162.

Kovach, R. L., Watkins, J. S. and Talwani, P., 1972. Active seismic experiment. In: *Apollo 16 Preliminary Science Report.* NASA SP-315, 10-1 to 10-14.

Kuiper, G. P., 1954. On the origin of the lunar surface features. *Proc. Natl. Acad. Sci. U. S., 40*: 1096.

Kuiper, G. P., Strom, R. G. and LePoole, R. S., 1966. Interpretation of the Ranger records. *J. P. L. Tech., Rep. 32-800*, 35-248.

Larimer, J. W. and Anders, E., 1967. Chemical fractions in meteorites – II. Abundance patterns and their interpretation. *Geochim. Cosmochim. Acta, 31*: 1239-1270.

Larimer, J. W. and Anders, E., 1970. Chemical fractions in meteorites – III. Major element fractions in chondrites. *Geochim. Cosmochim. Acta, 34*: 367-387.

Latham, G. V., Ewing, M., Press, F., and Sutton, G., 1969. The Apollo passive seismic experiment. *Science, 165*: 241-251.

Latham, G. V., Ewing, M., Press, F., Sutton, G., Dorman, J., Nakamura, Y., Toksoz, N., Lammlein, D. and Duennebier, F., 1972. Passive seismic experiment, In: *Apollo 16 Preliminary Sci. Rept.,* NASA SP-315, 9-1 to 9-29.

Laul, J. C., Morgan, J. W., Ganapathy, R., and Anders E., 1971. Meteoritic material in lunar samples: Characterization from trace elements. *Proc. Second Lunar Sci. Conf., Suppl. 2, Geochim. Cosmochim. Acta, 2*: 1139-1158.

Lindbald, B. A., 1967. Luminosity function of sporadic meteors and extrapolation of the influx rate to micrometeorite region. In: *Proceedings of a Symposium on Meteor Orbits and Dust.* NASA SP-135, 171-180.

Lindsay, J. F., 1972. Development of soil on the lunar surface. *Jour. Sediment. Petrology, 42*: 876-888.

Lindsay, J. F. and Srnka, L. J., 1975. Galactic dust lanes and lunar soil. *Nature, 157*:776-778.

Lindsay, J. F., Criswell, D. R., Criswell, T. L. and Criswell, B. S., 1975. Sound producing dune and beach sands. *Geol. Soc. Amer. Bull.* (in press).

Lovell, A. C. B., 1954. *Meteor Astronomy.* Claredon Press, Oxford, London, 463 pp.

Mandeville, J. C., and J. F. Vedder, 1971. Microcraters formed in glass by low density projectiles. *Earth Planet. Sci. Lett., 11*: 297-306.

Marcus, A. H., 1970. Comparison of equilibrium size distributions for lunar craters. *Jour. Geophys. Res., 75*: 4977-4984.

Mason, B. H., 1962. *Meteorites.* Wiley, N.Y., 274 pp.

Mason, B. H., 1971. *Handbook of Elemental Abundances in Meteorites.* Gordon and Breach, N.Y., 555 pp.

McCoy, J. E. and Criswell, D. R., 1974. Evidence for a high altitude distribution of lunar dust. *Proc. Fifth Lunar Sci. Conf., Suppl. 5, Geochim. Cosmochim. Acta, 3*: 2991-3005.

McCrosky, R. E. and Posen, A., 1961. Orbital elements of photographic meteors. *Smithson. Contrib. Astrophys., 4*: 15-84.

McDonnell, J. A. M., 1970. Review *in situ* measurements of cosmic dust particles in space. Paper presented at *13th COSPAR Meeting,* Leningrad.

McDonnell, J. A. M. and Ashworth, D. G., 1972. Erosion phenomena on the lunar surface and meteorites. In *Space Research XII*, Akademie-Verlag, Berlin, 333-347.

McDonnell, J. A. M., Ashworth, D. G., Flavill, R. P. and Jennison, R. C., 1974a. Simulated microscale erosion on the lunar surface by hypervelocity impact, solar wind sputtering, and thermal cycling. *Proc. Third Lunar Sci. Conf., Suppl. 3, Geochim Cosmochim. Acta, 3*: 2755-2765.

McDonnell, J. A. M., Flavill, R. P. and Ashworth, D. G., 1974b. Hypervelocity impact and solar wind erosion parameters from simulated measurements on Apollo samples. In: *Space Research—XIV*, Akademie-Verlag, Berlin, 733-737.

McDonnell, J. A. M. and Flavill, R. P., 1974. Solar wind sputtering on the lunar surface: Equilibrium crater densities related to past and present microparticle influx rates. *Proc. Fifth Lunar Sci. Conf., Suppl. 5, Geochim. Cosmochim. Acta, 3*: 2441-2449.

McKay, D. S., and Heiken, G. H., 1973. Petrography and scanning electron microscope study of Apollo 17 orange and black glass. *EOS, 54*: 599-600.

McKinley, D. W. R., 1961. *Meteor Science and Engineering.* McGraw-Hill, N.Y., 309 pp.

Morgan, J. W., Laul, J. C., Krahenbuhl, U., Ganapathy, R., and Anders, E., 1972. Major impacts on the moon: Characterization from trace elements in Apollo 12 and 14 samples. *Proc. Third Lunar Sci. Conf., Suppl. 3, Geochim. Cosmochim. Acta, 2*: 1377-1395.

Naumann R. J., 1966. The near earth meteoroid environment. NASA TN D-3711, 21 pp.

O'Keefe, J. A., 1969. Origin of the moon. *Jour. Geophys. Res., 74*: 2758-2767.

O'Keefe, J. A., 1970. The origin of the moon. *Jour. Geophys. Res., 75*: 6565-6574.

Öpik, E. J., 1958. On the catastrophic effects of collisions with celestial bodies. *Irish Astrophys. Jour., 5*: 34-36.

Öpik, E. J., 1966. The stray bodies in the solar system Part II. The cometary origin of meteorites. *Advanc. Astron. Astrophys., 4*: 301-336.

Plavec, M. 1955. Meteor stream at early stages of evolution from meteors. T. R. Kaiser (Editor), *Meteors.* Pergamon, N.Y., 167-176.

Reid, A. M. Ridley, W. I., Donaldson, C. and Brown, R. W., 1973. Glass compositions in the orange and gray soils from Shorty Crater, Apollo 17. *EOS, 54*: 607-609.

Rennilson, J. J. and Criswell, D. R., 1974. Surveyor observations of lunar horizon-glow. *The Moon, 10*: 121-142.

Ringwood, A. E., 1966. Chemical evolution of the terrestrial planets. *Geochim. Cosmochim. Acta, 30*: 41-104.

Ringwood, A. E., 1970. Petrogenesis of Apollo 11 basalts and implications for lunar origin. *Jour. Geophys. Res., 75*: 6453-6479.

Roosen, R. G. 1970. The Gegenschein and interplanetary dust outside the earth's orbit. *Icarus, 13*: 184-201.

Schneider, G., Störzer, D., Mehl, A., Hartung, J. B., Fechtig, H. and Gentner, 1973. Microcraters on Apollo 15 and 16 samples and corresponding cosmic dust fluxes. *Proc. Fourth Lunar Sci. Conf., Suppl. 4, Geochim, Cosmochim. Acta, 3*: 3277-3290.

Shoemaker, E. M., Hackman, R. J. and Eggleton, R. E., 1962. Interplanetary correlation of geologic time. *Advan. Astronaut. Sci., 8*: 70-89.

Shoemaker, E. M., Hait, M. H., Swann, G. A., Schleicher, D. L., Schaber, G. G., Sutton, R. L. and Dahlem, D. H., 1970. Origin of the lunar regolith at Tranquility Base. *Proc. Apollo 11 Lunar Sci. Conf., Suppl 1, Geochim. Cosmochim. Acta, 3*: 2399-2412.

Siedentopf, H. 1955. Zur Optischen Deutung Des Gegenscheins. *Z. Astrophys. 38*: 240-244.

Silver, L. T., 1970. Uranium-thorium-lead isotopes in some Tranquility Base samples and their implications for lunar history. *Proc. Apollo 11 Lunar Sci. Conf., Suppl. 1, Geochim. Cosmochim. Acta, 2*: 1533-1574.

Singer, S. F. and Walker, E. H., 1962. Electrostatic dust transport on the lunar surface. *Icarus, 1*: 112-120.

Soberman, R. K., 1971. The terrestrial influx of small meteoric particles. *Rev. Geophys. Space Phys., 9*: 239-244.

Southworth, R. B., 1967a. Phase function of the zodiacal cloud. In: J. L. Weinberg (Editor), *Proc. Symp. Zodiacal Light Interplanet. Medium.* NASA SP-150, 257-270.

Southworth, R. B., 1967b. Space density of radio meteors. In: J. L. Weinberg, (Editor), *Proc. Symp. Zodiacal Light Interplanet. Medium.* NASA SP-150, 179-188.

Swann, G. A., and others, 1972. Preliminary geologic investigation of the Apollo 15 landing site. In: *Apollo 15 Preliminary Science Report.* NASA SP-289, 5-1 to 5-112.

Tera, F. and Wasserberg, G. J., 1974. U-Th-Pb Systematics on lunar rocks and inferences about lunar evolution and the age of the moon. *Proc. Fifth Lunar Sci. Conf., Suppl. 5, Geochim. Cosmochim. Acta, 2:* 1571-1599.

Urey, H. C., 1952. *The Planets; Their Origin and Development.* Yale, New Haven, 245 pp.

Urey, H. C. and MacDonald, G. J. F., 1971. Origin and history of the moon. In: Z. Kopal (Editor), *Physics and Astronomy of the Moon,* Academic Press, 213-289.

Van de Hulst, H. C., 1947. Zodiacal light in the solar corona. *Astrophys, Jour., 105:* 471-488.

Vedder, J. F., 1966. Minor objects in the solar system. *Space Sci. Revs., 6:* 365-414.

Vedder, J. F., 1971. Microcraters in glass and minerals. *Earth Planet. Sci. Lett., 11:* 291-296.

Whipple, F. L., 1950. A comet model. I. The acceleration of the comet Encke. *Astrophys. Jour., i11:* 365-395.

Whipple, F. L., 1951. A comet model. II. Physical relations for comets and meteors. *Astrophys. Jour., 113:* 464-474.

Whipple, F. L., 1963. On the structures of the cometary nucleus. In: B. M. Middlehurst and G. P. Kuiper (Editors), *The Moon Meteorites and Comets,* Univ. Chicago Press, Chicago, 639-664.

Whipple, F. L., 1967. On maintaining the meteoritic complex: In: J. L. Weinberg (Editors), *Zodiacal Light and the Interplanetary Medium.* NASA SP-150, 409-426.

Whipple, F. L., 1967. On maintaining the Meteoritic complex. *Smithson. Astrophys. Obs. Spec. Rept. No. 239.* 1-46.

Wise, D. U., 1963. An origin of the moon by rotational fission during formation of the earth's core. *Jour. Geophys. Res., 68:* 1547-1554.

Wood, J. A., 1962. Chondrules and the origin of the terrestrial planets. *Nature, 194:* 127-130.

CHAPTER 3

Hypervelocity Impact

Today when we view the densely cratered surface of the moon we see it as clear evidence of the importance of the meteoroid flux in determining the moon's morphology. From Chapter 2 it is apparent that most of the erosive energy available at the lunar surface is meteoritic in origin. While the meteoroid flux is a tenuous energy source by terrestrial standards, the lack of an atmosphere and the long periods of time available coupled with the extremely high velocities of individual impacts makes it an effective lunar sedimentary process. However clear this may seen today, the largely meteoritic origin of lunar craters has only recently been generally agreed to.

The origins of the impact hypothesis are to some extent obscure. Green (1965) attributes the first mention of the hypothesis to von Beberstein in 1802 although others place the first mention with Bruitheisen in 1829. Certainly by 1893 Gilbert had laid out very clear arguments for the impact origin of the large circular mare such as Mare Imbrium. Paralleling the impact hypothesis was the volcanic hypothesis which was given its initial boost in 1846 by Dana. Numerous publications in favor of both the volcanic hypothesis (for example Spurr, 1944-1949; Green, 1965, 1970, 1971) and the impact hypothesis (for example Baldwin, 1949; Urey, 1952; Shoemaker, 1962) appeared with increasing frequency into the late 1950's and early 1960's.

The length of time required to resolve the problem of lunar crater origins reflects failings more on the part of the impact school than the volcanic school of thought. Much of the discussion, beginning with Gilbert (1893), revolved around the circularity of the craters. If the craters were of impact origin why were they so circular; any variations in the angle of incidence of the projectile would cause the craters to have different shapes. Volcanic activity could be observed on the earth first hand. However, it was not until quite recently that hypervelocity impact experiments in the laboratory showed that impacts at orbital velocities resulted in what could be described (although incorrectly) as explosions which produced relatively circular craters independent of the angle of incidence of the projectile (Shoemaker, 1962).

By the mid 1960's Baldwin (1965) felt confident enough about the impact hypothesis to suggest that the long standing battle was over — most people in the field now share this view. However, the arguments are still, even after the lunar landings, based largely on crater morphology and as late as 1971 the Lunar Science Institute in Houston hosted a meeting entitled "Meteorite Impact and Volcanism." Some idea of the complexity of the morphological identification of impact versus volcanic craters can perhaps be gained from the recent study of Pike (1974). He treated the question from a multivariate view point and concluded that no single morphologic parameter separated the two cratering modes with any degree of certainty. Even in the multivariate analysis maars and tuff rings could be grouped morphologically with other craters of either impact or volcanic genesis. While it seems clear enough today that most of the lunar craters are of impact origin it seems that at least a modest proportion of lunar craters should be volcanic. Their positive identification may have to wait for further lunar exploration.

Cratering Mechanics

The collision of a large meteoroid with the lunar surface offers perhaps the ultimate example of hypervelocity impact in the solar system. On a time scale of minutes large geologic structures the magnitude of the lunar mare are created (Gault and Heitowit, 1963). The energy expended in the formation of these large structures is orders of magnitude larger than the largest nuclear explosion. Because of the violence of these events they are frequently compared to explosions. The comparison, while offering a useful analogy, is not strictly correct. Hypervelocity impacts can produce very high specific internal energies. These energies are, however, the consequence of the mechanical compression of the projectile as it penetrates the target, rather than the cause of the compression. An impact, accompanied by vaporization, is an explosion only to the extent that debris is thrown upward and outward from a transient cavity.

The sequence of events leading to the formation of an impact crater can, for convenience, be divided into three separate stages which are determined by the physical processes taking place and the associated time scale involved in their completion (Gault *et al.,* 1968). These stages are (1) the compression stage, (2) the excavation stage and (3) the modification stage.

Compression Stage

At the moment of contact between the projectile and the target a system of shock waves is established that provides the mechanism for transferring the kinetic energy of the impacting projectile into the target material. Two shock fronts are produced, one in the projectile, the other in the target. One shock wave runs upward into the projectile the second downward into the target. Initially the shock-compressed material is limited to a small lens-shaped mass directly in front of the projectile (Shaded area in Figure 3.1). Particle motion in this initial lens shaped zone is predominently downward. The pressures at this stage are of the order of a few megabars. Since most natural materials have strengths of less than a few kilobars the deformational stresses are 10^3 to 10^4 times greater than the material strength. Effectively the compression stage is a hydro-dynamic or fluid-flow phase of cratering (Charters, 1960).

The shock waves engulf an ever increasing mass of target and projectile as the projectile penetrates more deeply into the target. The shock wave system is modified dramatically with time due to the free surfaces on the faces of the target and projectile. Since a free surface cannot sustain a state of stress the shock waves race outward along the face of the target and upward along the sides of the projectile. Consequently a family of rarefraction waves develops behind the shock waves as a means of decompressing the high pressure zone behind the shock waves. That is, the initial lens-shaped zone enclosed between the two shock waves is no longer able to maintain its integrity and it essentially opens up around its edges. The appearance of rarefraction waves immediately precedes the onset of jetting (Fig. 3.1), the hydrodynamic ejection of mass at extreme velocities (Gault et al., 1963). Jetting velocities are multiples of the impact velocity (Fig. 3.2). The jetted materials have been subjected to the highest pressures and hence highest temperatures of any materials ejected by the impact and are in a liquid or vapor state, depending on the impact velocity.

The compression stage is terminated when the shock wave in the projectile reflects from the back of the projectile and "terminal engulfment" (as Gault et al., 1968 call the process) occurs (Fig 3.1). In laboratory experiments this phase is completed in less than 10^{-6} s but for larger bodies up to 1 km diameter as much as 10^{-1} s would be required. Once the projectile is consumed by the shock waves the cratering process is essentially one of gradual stress relaxation from the initial projectile contact. The stress wave geometry is very complex but is dominated by a spherically expanding shell of compressed target and projectile material. The free surfaces and associated rarefraction waves cause the shocked material at, and just below, the surface to begin to deflect laterally outward and upward to initiate the excavation stage.

Compression Stage

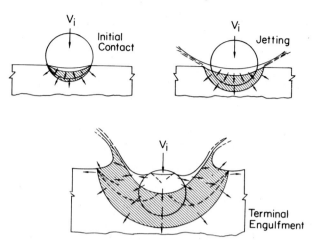

Fig. 3.1. Schematic representation of the compression stage of the formation of an impact crater (from Gault *et al.*, 1968).

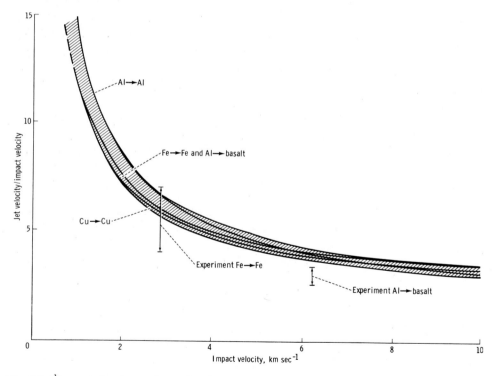

Fig. 3.2. Theoretical and experimentally observed jetting velocities for impact in metals as a function of the impact velocity (from Gault *et al.*, 1968).

Excavation Stage

The basic shock wave geometry during the excavation stage is an approximately hemispherical shell of compressed material, expanding radially and distributing the kinetic energy over a steadily increasing mass of target material (Fig. 3.1). Compared to the jetting phase, pressures and ejection velocities are low but by far the greatest mass of material is removed during this stage. The expanding shock front moves with a velocity V_s and imparts a particle motion of U_p to the target material. Since the total energy is constant the energy density in the shocked material decreases as the wave front expands and some energy is irreversibly converted to other forms such as heat. The radial movement of shock waves along the face of the target is accompanied by a fan of rarefaction waves (Fig. 3.3) which maintain the zero stress boundary condition at the free surfaces. A stress wave pattern is thus established that diverts the motion of the shocked material away from the radial direction imparted initially by the shock wave. From Figure 3.3 it can be seen that only the velocity component of the particle motion normal to an isobar is accelerated as it passes to a lower pressure. Consequently particle motion is continually deflected by a series of incremental velocity changes ΔU_r towards the direction of maximum pressure gradient. The total velocity increment U_r produced by all the expansion waves is added vectorially to U_p yielding a resultant particle velocity U_e (Gault *et al.*, 1968). The movement of near surface particles is consequently almost horizontal whereas deeper materials are driven downwards and then reflected upwards. (Fig. 3.4).

Crater growth is a well regulated process. During most of the growth

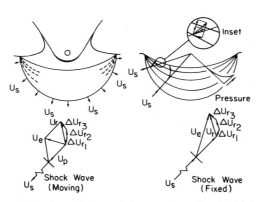

Fig. 3.3. Schematic representation of the radial expansion of material behind a shock wave by a hypervelocity impact (from Gault *et al.*, 1968).

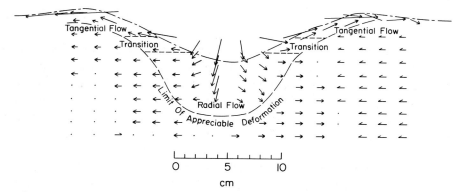

Fig. 3.4. Flow diagram showing total movement produced by the cratering process in individual point masses (from Gault *et al.*, 1968).

period the enlarging cavity maintains a shallow basin-like geometry which is essentially the same as the final crater form. The ejecta move up and shear out from the upper wall of the growing cavity in a steady flow that develops into an inverted conical ejecta curtain. The ejecta curtain then quickly tears apart into a delicate filamentory pattern which is consequently reflected in the ray patterns so common on the lunar surface.

Crater formation is completed in a time interval of the order of 10^5 times that of the compression stage. The transport of the ejecta beyond the crater can take much longer and is an extremely complex process. Later sections deal with this phase in detail.

Modification Stage

The principle post-cratering modification processes include (1) slumping of the peripheral rim structure in the impact basin (2) filling and possible formation of a central dome of isostatic readjustment (3) erosion and resultant infilling (4) filling with material from internal sources — for example flooding by lava. These processess are all discusses in varied detail in other sections.

Crater Shape

A knowledge of the shape of craters is essential in understanding the stratigraphy of the moon and in establishing models for erosion and transportation of sedimentary materials on the lunar surface. Shape parameters have also provided the basis for much of the discussion concerning the origin of lunar craters. The relationships among the various parameters are complex and among other things are scale dependent — despite this fact most analyses use normalized parameters for comparative purposes. Because of the scale dependency it is convenient to treat craters in three rough groupings determined by scale; (1) microcraters or pits, (2) macrocraters, (3) megacraters. The division between the three groups is somewhat arbitrary in that it depends largely on the projectile mass-range studied experimentally.

Microcraters

Glass-lined microcraters, or "pits" as they are frequently called, are very abundant on the surfaces of most lunar rocks and are clearly important in the erosional cycle. Little is known of the effects of hypervelocity impact of very small particles into loose detrital materials but it has been suggested that soil particles called agglutinates which consist of an intimate mixture of glass and soil are the result of single micrometeoroid impacts (Lindsay, 1975; Lindsay and Srnka, 1975).

The morphology of microcraters produced by particles with masses between 10^{-11} and 10^{-13} g in glass substrates varies considerably, particularly in response to changes in velocity (Mandeville and Vedder, 1971). At low velocities (≈ 3 km s^{-1}) the rebounding projectile leaves a shallow depression in the target. At 3.5 to 4.6 km s^{-1} a continuous lip overflows the rim of the depression. At 5 km s^{-1} the lip becomes petalled in form and a spallation zone appears for the first time. The spallation zone lies just outside the central pit and is characterized by radial and concentric fractures and radial ridges where the spalls have been ejected. As the velocity is further increased the spallation zone becomes more finely fragmented.

At a constant velocity the ratio of the pit diameter to the projectile diameter (Dp/d) is independent of mass but as the velocity increases the ratio also increases:

$$Dp/d = 0.76 \, V^{0.32} \tag{3.1}$$

where V is velocity in km s^{-1}. Similarly for a constant velocity the diameter of the central pit (Dp) varies approximately with the cube root of the mass of the projectile (Mandeville and Vedder, 1971). The diameter of the spall zone (Ds) at a constant velocity likewise increases with increasing mass but the exponent is slightly larger indicating that the diameter of the spall zone is increasing faster than the diameter of the pit (Dp):

$$Dp = 1.6 \, m^{0.33} \qquad Ds = 2.9 \, m^{0.42} \qquad 5.3 < V < 6.4 \text{ km s}^{-1} \tag{3.2}$$

$$Dp = 1.8\ m^{0.33} \qquad Ds = 3.6\ m^{0.38} \qquad 7.0 < V < 9.0\ km\ s^{-1} \tag{3.3}$$
$$Dp = 2.1\ m^{0.27} \qquad Ds = 4.2\ m^{0.39} \qquad 12.0 < V < 14.0\ km\ s^{-1} \tag{3.4}$$

where V is in km s^{-1}, Dp and Ds are in micrometers and m is in picograms (1 pg = 10^{-12} g).

Crater morphology is controlled by the angle of incidence (i) of the projectile so that the craters become elongate and shallow and the spallation threshold occurs at higher velocity. At i = 45° spallation begins at 6.5 km s^{-1} compared to 5.2 km s^{-1} at a normal angle of incidence. When the angle is further reduced to i = 30° a velocity of 12.3 km s^{-1} is required to initiate spallation. Craters formed at i = 30° were shallower and more elongate than those formed at i = 45°. The total mass of damaged material including the spall zone (Mes) is a direct function of kinetic energy (KE):

$$Mes = 230\ (KE)^{1.1} \tag{3.5}$$

as is the mass ejected from the central pit (Mec):

$$Mec = 47\ (KE)^{1.1} \tag{3.6}$$

where Mes and Mec are in picograms and KE is in microjoules (Mandeville and Vedder, 1971). The fact that in both cases the exponent is larger than 1 indicates that cratering effeciency increases with increasing energy which is a function of both mass and velocity. Similar relationships appear to exist for macrocraters in crystalline rock (Gault, 1973).

Data are available on the effects of microcratering in loose particulate substrates comparable to the lunar soil. The data are limited to impacts involving energies from approximately 0.14 ergs to 50 ergs within a velocity range of 2.5 to 12 km s^{-1} (Vedder, 1972). For a weakly cohesive substrate with a maximum grain size comparable to the projectile diameter, the crater produced is approximately 25 times the diameter of the projectile and detrital materials of approximately 3 to 4 orders of magnitude greater than the projectile mass are displaced by ejection and compaction. In terms of energy approximately 1 microjoule is required to excavate 1 microgram of loose detrital materials. Because of the scatter of the data it is not possible to establish accurately the relationships among parameters. However, the craters are bowl-shaped and have no raised rim. When the grain size of the substrate is increased to an order of magnitude larger than the diameter of the projectile the craters are still recognizable as such. The area of disturbance is comparable to impacts in finer detrital materials but the diameter and depth of the craters are not clearly defined due to the effects of single grain displacement. A ratio of 1:10 for projectile to particulate grain size appears to mark the transition from primarily multigrain interaction to a single grain impact. If the grain size of the particulate material is increased further microcraters are formed on the surface of individual particles.

Craters formed by particles with masses less than 10^{-5} g impacting cohesionless particulate material produce craters with a diameter (Da) proportional to KE:

$$Da = 6 \times 10^{-3} (KE)^{0.352} (Sin\ i)^{1/3} \qquad (3.7)$$

where i is the angle of incidence in relation to the local horizontal (cgs units) (Gault, 1974). The exponent associated with the kinetic energy is intermediate to the exponent for the pit and spall diameters in equations 3.2 to 3.4 suggesting a similar partitioning of energy.

Macrocraters

Hypervelocity impact events which form macrocraters are important agents in the development of the lunar soil as they erode bedrock and rework the soil. The shape of macrocraters formed in bedrock is relatively well known. Some data are available for impacts into loose particulate material but they are limited.

The most complete study of macrocraters into bedrock is that of Gault (1973) who has determined empirically from experimental impacts the relationship between the ejected mass (Me), depth (d), diameter (Da) and angle of incidence (i) in relation to the kinetic energy of the event. The relationships are believed to be valid for impact kinetic energies within the range 10 to 10^{16} ergs (Gault, 1973). This is equivalent to a particle mass range of 5×10^{-12} to 5×10^{3} g if a root-mean-square velocity of 20 km s^{-1} is assumed, or to craters with diameters of 10^{-3} to 10^{3} cm.

The mass ejected (Me) from a macrocrater can be determined by:

$$Me = 10^{-10.061} \left[\rho_p / \rho_t \right]^{1/2} (KE)^{1.133} (Sin\ i)^2 \qquad (3.8)$$

where ρ_p and ρ_t are the density of the projectile and target respectively, and KE is the kinetic energy ($\frac{1}{2}m\ V^2$) of the event if m and V are the projectile mass and velocity respectively. This relationship is very similar to equation 3.5 for microcraters but produces smaller values for Me.

The exponent associated with the kinetic energy is greater than 1.0. That is, the ejected mass is increasing with event magnitude. This is attributable to a change in the effective target strength in relation to the dimensions of the masses of the target material which are engulfed and deformed by the stress waves produced during impact (Moore et al., 1965). The value of the exponent appears to depend upon the mode of failure of the material which is fragmented and ejected from the crater. Failure solely in tension occurs with an exponent of 1.2 while compressive failure produces a value of 1.091. The observed intermediate value suggests that both mechanisms are important but that tensile failure is dominant (Gault, 1973; Moore et al., 1961, 1965).

The depth (p) of a macrocrater in crystalline rock is determined by

$$p = 10^{-3.450} \, \rho_p^{1/6} \, \rho_t^{-\frac{1}{2}} \, (KE)^{0.357} \, (Sin \, i)^{0.66} \qquad (3.9)$$

The exponent associated with kinetic energy is considerably less than 1. Thus crater depth decreases with event magnitude.

Crater diameter (Da) changes in much the same manner as depth

$$Da = 10^{-2.823} \, \rho_p^{1/6} \, \rho_t^{-\frac{1}{2}} \, (KE)^{0.370} \, (Sin \, i)^{0.86} \qquad (3.10)$$

Within the size range considered craters should have similar diameter to depth ratios with a minor dependence upon Sin i and KE:

$$Da/p = 10^{0.617} \, (KE)^{0.013} \, (Sin \, i)^{0.2} \qquad (3.11)$$

Craters are generally relatively circular in plan but spalling of large pieces causes the outline to be erratic; consequently Da is the average diameter.

The formation of macrocraters in cohesionless substrates has received more attention than microcrater formation in loose substrates but the date are nowhere near as complete as for impacts into crystalline rock materials.

The diameter (Da) of a macrocrater produced by a meteoroid larger than 10^{-5} g can be related empirically to kinetic energy (KE) (Gault et al., 1974) by:

$$Da = 2 \times 10^{-2} \, (KE)^{0.280} \, (Sin \, i)^{1/3} \qquad (3.12)$$

The value of the energy exponent is smaller than for microcraters in loose particulate materials (see equation 3.7), possibly indicating that cratering efficiency is decreasing rapidly with event magnitude.

Macrocraters formed in non-cohesive particulate materials are basin shaped and could be will represented by a spherical segment (Fig. 3.5). The

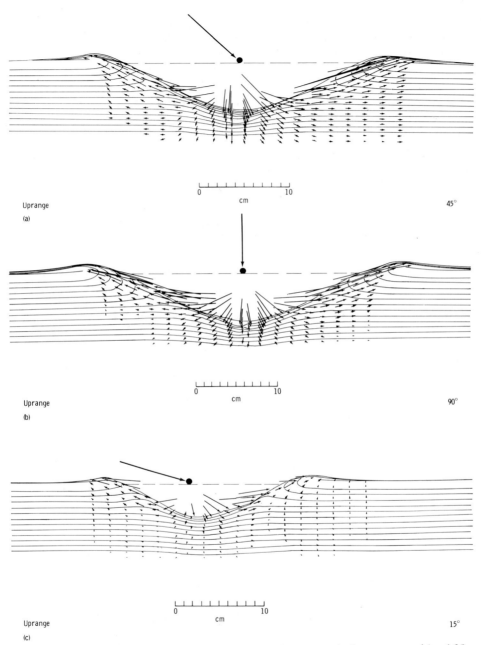

Fig. 3.5. Mass-vector displacements produced by impacts into non-cohesive quartz sand by 6.35 mm aluminum spheres: (a) Trajectory angle, 45°; (b) Trajectory angle, 90°; impact velocities 6.4 and 6.8 km s⁻¹, respectively; (c) Trajectory angle, 15°; impact velocities, 6.8 km s⁻¹ (from Gault *et al.*, 1974, *Proc. 5th Lunar Sci. Conf., Suppl. 5, Vol. 3*, Pergamon).

macrocraters have well-developed rims in contrast to microcraters. Strong horizontal thrusting occurs in the crater walls at about middle depths. At the bottom of the crater horizontal strata are displaced radially downwards and thinned, partially due to radial displacement but primarily as a result of compaction, a variable of little importance in impacts into crystalline rock. Approximately half of the crater depth ($\approx$15 per cent by volume) is due to compaction independently of the angle of incidence of the impacting projectile (Fig. 3.5).

Meteoroids large enough to produce macrocraters are probably the main means by which the lunar soil grows in thickness. When a meteoroid penetrates the blanket of lunar soil and interacts with the bedrock the morphology of the craters is determined by the ratio (R) of the crater rim diameter to the thickness of the soil layer (Oberbeck and Quaide, 1967; Quaide and Oberbeck, 1968). When R is less than approximately 4.25 normal crater development occurs because 4.25 is close to the normal diameter to depth ratio for craters formed in loose substrates (Fig 3.6). When R is increased to between 4.25 and 9.25 the bedrock substrate

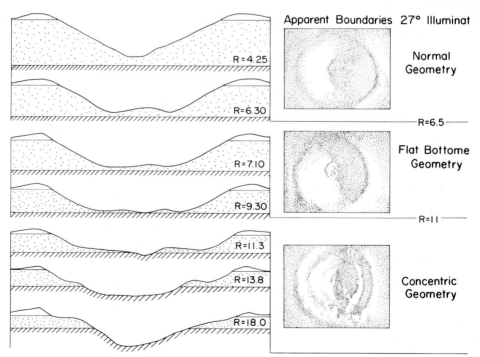

Fig. 3.6. The relationship between crater morphology and depth to bedrock beneath a noncohesive soil layer. R is the ratio of crater rim to the soil thickness (after Oberbeck, V. R. and Quaide, W. L., *Jour. Geophys. Res.*, 72: 4697-4704, 1967, copyrighted American Geophysical Union).

interferes with normal crater development and flat bottomed craters result. When R is about 6.3 a central mound may develop in the bottom of the crater. At values of R larger than 9.25 concentric craters characterized by a double annular or arcuate structure are formed. These craters are essentially complex crater-within-crater structures. Knowledge of these parameters has provided considerable information on the thickness of the lunar soil and on the growth of the soil layer.

Megacraters

This group includes all of the craters visible in orbital photographs up to and including the giant craters of the circular mare. In terms of scaling problems this group of craters is complex and should probably be dealt with in at least two subgroupings. As the scale of these larger events increases the craters cease to be simple scars on the planetary surface as they begin to interact with the planetary processes themselves. For example, a considerable amount of isostatic readjustment must occur in larger craters and in craters of the size of Mare Imbrium the planetary heat flow balance is disturbed, possibly resulting in igneous activity.

Baldwin (1963) made the most comprehensive study of large scale cratering problems and has provided the basis for most of the following discussion. Baldwin's (1963) empirical relationships are based on parameters determined from both lunar and terrestrial impact craters in relation to experimental explosion craters where the geometry and energy of the explosion are well known. The main difficulty in using experimental explosions in the energy calibration of impact craters is due to "Lampson scaling" effects as Baldwin (1963, p. 130) has called them. The shape and size of an explosion crater of known energy (W) is determined to a large extent by the effective use of the available energy which, in turn, is largely determined by the height or the depth of the burst center above or below the ground surface. The main parameter of concern is the scaled depth-of-burst $H/W^{1/3}$. Baldwin (1963) went to considerable length to establish the relationship between the scaled depth-of-burst of explosions and impacts and found that to a large extent it is determined by the size of the impacting projectile. From this he established a series of empirical relationships to be used for craters within certain size ranges based on crater diameter.

In contrast to studies of micro- and macrocraters Baldwin has found it necessary to fit higher order polynomials to the log-log data to obtain a satisfactory fit. Crater diameters are in kilometers, apparent crater depths and rim heights in meters, all energy in in ergs.

The diameter (Dl) of a crater relates to the kinetic energy (El) of the impacting projectile in the following manner:

$$D_l = -7.2089 + 0.304345\ E_l + 1.733 \times 10^{-3}\ E_l{}^2 - 7.039 \times 10^{-5}\ E_l{}^2$$
$$+2.9 \times 10^{-7}\ E_l{}^4 \qquad D_l < 3\ \mathrm{km} \qquad\qquad\qquad (3.13)$$

$$D_l = -7.1293 + 0.284136\ E_l + 2.866 \times 10^{-3}\ E_l{}^2 - 0.590 \times 10^{-5}\ E_l{}^3$$
$$+ 4.3 \times 10^{-7}\ E_l{}^4 \qquad 3 < D_l < 7\ \mathrm{km} \qquad\qquad\qquad (3.14)$$

$$D_l = -7.9240 + 0.3284\ E_l \qquad D_l > 7\ \mathrm{km} \qquad\qquad\qquad (3.15)$$

It is perhaps not necessary to make the point but the higher order terms in equations (3.13) and (3.14) are all small and for most applications they can be neglected. The first order coefficients of these equations are all smaller than the equivalent exponent for macrocraters. That is, the diameter of large scale craters does not increase as rapidly with increasing event magnitude as it does for macrocraters. Likewise microcraters have a larger exponent again. This suggests that if a general relationship is to be established between crater diameter and kinetic energy, the exponent must be treated as a function of event magnitude. Thus, despite the fact that the higher order coefficients in equations 3.13 and 3.14 are small, the approach used by Baldwin (1963) is probably the only means of establishing more general empirical relationships.

Apparent crater depth (pl) is the second most important shape parameter to be determined for craters, and like diameter it can be related directly to kinetic energy:

$$p_l = -5.4649 + 0.40763\ E_l - 3.375 \times 10^{-3}\ E_l{}^2 \quad D_l < 3\ \mathrm{km} \qquad (3.16)$$

$$p_l = -5.444 + 0.4136\ E_l - 3.675 \times 10^{-3}\ E_l{}^2 \quad D_l > 3\ \mathrm{km} \qquad (3.17)$$

Again the higher order terms are relatively small as are the diferences between the two equations. When converted to equivalent units the intercept of the above two equations is similar to equation (3.9) for macrocraters, however, the first order term for the above equations is larger than the equivalent exponent of equation (3.9). As with the diameter it seems that rates of change of crater depth are controlled by event magnitude as well.

Megacraters are the origin of much of the large scale lunar stratigraphy. It is therefore useful to know the rim height of larger craters to gain some insight into stratigraphic relations and thinning of ejecta blankets. This latter problem is discussed in some detail in a following section of this chapter.

The logarithmic rim height (R$_l$) can be related to D$_l$ and thus E$_l$ in the following way (Baldwin, 1963):

$$R_l = 1.5987 + 0.9098 \ D_l - 8.506 \times 10^3 \ D_l{}^2 + 4.366 \times 10^{-3} \ D_l{}^3$$
$$D_l < 3 \ \text{km} \qquad (3.18)$$

$$R_l = 1.4847 + 0.9098 \ D_l - 8.506 \times 10^{-3} \ D_l{}^2 + 4.366 \times 10^{-3} \ D_l{}^3$$
$$D_l > 3 \ \text{km} \qquad (3.19)$$

The pressure behind the shock front produced by an impacting meteoroid persists at high levels relative to the strength of the substrate for a considerable distance into the bedrock. Much of the substrate is subjected to stresses of sufficient magnitude to cause fracturing and fragmentation. Consequently there is a lens shaped zone beneath an impact crater which is brecciated and fractured by the impact but not ejected. The lens of fractured materials fades in all directions as the fractures become further apart (Baldwin, 1963). Baldwin found that the depth to the limit of major brecciation below the original ground level increased at a greater rate than the true depth of the crater below the original ground level (apparent crater depth minus rim height). The thickness of the breccia lens (B) in meters is

$$B = 0.624 \ D_t{}^{1.2658} \qquad (3.20)$$

where D$_t$ is the true crater depth in meters. The breccia lens is essentially non-existent beneath craters with diameters smaller than 35 m. It follows from the above equation that the outer crust of the moon should be shattered to a considerable depth. Seismic data indicate that the outer kilometer of the crust has extremely low seismic velocities (<1 km s^{-1}) which are consistent with a zone of shattered rock (Nakamura $et\ al.$, 1974).

Energy Partitioning During Impact

Partitioning of energy during a hypervelocity impact is complex. It is determined not only by the more obvious variables such as the velocity of the meteoroid, but also by the nature of the impacted substrate and the magnitude of the event. In this section I wish to look not only at the proportionate partitioning of energy, but also at the way in which energy contributes to the production of clastic materials and the way in which they are transported and incorporated into the lunar stratigraphic record. The data used in the following discussion are varied, largely because of the difficulties in simulating hypervelocity impacts. Much of the data are derived from small scale experiments and it must be assumed that they can be scaled to larger events; other data came from studies of large explosion craters which are similar to, but not necessarily the same as, craters produced by impact.

There are four obvious ways in which the kinetic energy of the impacting projectile is expended (1) heating, (2) compacting, (3) comminuting and (4) ejecting detrital materials (Table 3.1).

Impact Heating

A relatively large proportion of the kinetic energy of an impact is irreversibly lost as heat. Some of this heat is absorbed by the projectile and a larger proportion by the target. Heat energy can be put to use in three ways (1) vaporization (2) fusion or (3) it can be effectively lost from the system by radiation or conduction. Most of the phase changes due to heating probably occur close to the time when jetting is initiated as the wave fronts in the projectile and target begin to separate and release pressure.

The heat absorbed by the projectile is noticeably higher in impacts into solid substrates (e.g. basalt). This occurs because the impact pressures are at least 50 per cent higher in a noncompressable target. Vaporization of the projectile should thus be more complete in bedrock impacts than impacts into less competent substrates. Conversely the heat released into the target during impact is higher in a loose substrate than a solid substrate. Melting and vaporization should thus be much more important for small impacts into lunar soils.

Gault *et al.* (1972) approached the problem of the partitioning of heat

TABLE 3.1

Partitioning of kinetic energy during hypervelocity impact (after Braslau, 1970).

	Energy Expended (per cent)	
	Sand Target	Basalt Target
Waste Heat		
Projectile	6	4-12
Target	26	19-23
Compaction	20	1
Comminution	8	10-24
Ejection	53	42-53
TOTAL	113	77-113

energy from a theoretical view point. They assumed that energy deposited irreversibly behind the shock front is trapped and stored as an increase in internal energy in the target material. The increase in internal energy, which is assumed to be manifested as heat energy for fusion and vaporization, is evaluated using the Hugoniot curve as an approximation for the release adiabat to ambient pressure from local peak shock pressure. In figure 3.7 the results of the calculations simulating the impact of a stony meteoroid into a tuff are presented. These estimates are probably all high as the effects of deformation and comminution are not considered. All of these effects are included as shock heating. The dependence upon velocity is clear. Below a velocity of 10 km s⁻¹ nothing is vaporized and below 5 km s⁻¹ there is no fusion. Since most experimental studies are made at velocities considerably less than 10 km s⁻¹ the difficulties of studying impact melts are obvious.

The proportion of heat energy involved in the production of impact melts on the moon is not known. Considerable work has, however, been done on the amount of melts produces in terrestrial impact craters (Dence, 1964, 1965, 1968, 1971; Short, 1965, 1970; Beals, 1965; Head, 1974; French, 1970; Engelhardt, 1972). In general impact melts occur as two facies in terrestrial craters (1) as a crystalline matrix of breccias and as glassy fragments in the breccias and (2) as discrete layers of melted rock which are completely recrystallized and resemble igneous rocks. In small simple

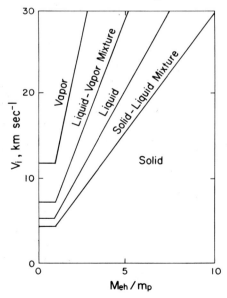

Fig. 3.7. Calculated variation with impact velocity Vi of the shock heated masses Meh in a target having a density of 1.7 g cm3 (tuff) normalized with the projectile mass mp whose density is 2.86 g cm3 (basalt) (after Gault *et al.*, 1972, *Proc. 3rd Lunar Sci. Conf., Suppl. 3, Vol. 3*, MIT/Pergamon).

terrestrial craters most of the melt occurs in a puddle at the bottom of the crater (Dence, 1968). Astronaut observation on the lunar surface suggest a similar distribution for small craters (<1 m) in the lunar soil. In larger more complex terrestrial craters the melt layer forms an annular ring between the central uplift and the rim which can be as much as several meters thick (Dence, 1965, 1968; French, 1970). The results of the terrestrial impact crater studies suggests that the melt volume (Vm) is related to crater diameter (D_a) as follows (Dence, 1965):

$$Vm = 0.002 \, D_a{}^3 \qquad\qquad\qquad (3.21)$$

where Da is in km (Fig. 3.8). Clearly this relationship can only be a first order approximation because energy will be partitioned according to impact velocity as well as to total event energy. Head (1974) has used this relationship to estimate the volume of impact melt in Mare Orientale as 75,000 km³ of internally-shock-melted material and a further 180,000 km³ of mixed breccia and melt. Orientale has a diameter of 620 km. The impact-melt origin of the materials forming the floor of Orientale is not obvious. However, some large craters such as Tycho (Fig. 3.9) have an extremely rough floor with abundant evidence of fluid flow.

Compaction

Little is known about the effects of compaction by hypervelocity impact. The density of lunar soil (which is dealt with in some detail in Chapter 6) does increase with depth suggesting that compaction is at least potentially important. Intuitively one would expect the effects of compaction to be greater in loose materials (such as the lunar soil) than a solid rock simply because of porosity. Braslau (1970) found that energy loss due to compaction in loose materials was considerable (20 per cent). A volume equivalent to 8 per cent of the material that could have been expected as ejecta from craters in solid rock was lost to compaction in loose materials.

Comminution

Although the strength of the shock wave propogating from the point of impact decays rapidly below the level commensurate with a deposition of a significant amount of irreversible heat energy, the pressure behind the shock front nevertheless persists at a high level relative to the strength of the target materials for a considerable distance into the target. Ultimately the shock wave decays into elastic waves but not before an appreciable mass is

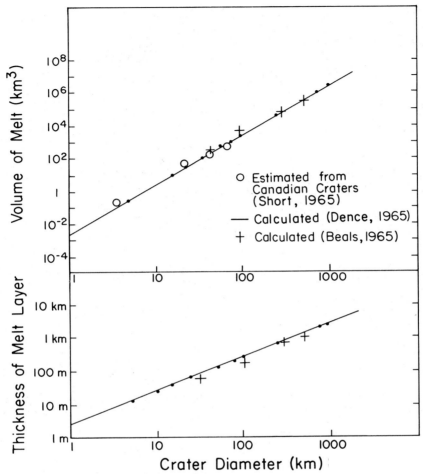

Fig. 3.8. Relationship between crater diameter and amount of impact melt formed by direct fusion of target rock. The regression line is based on Canadian craters (after Head, 1974).

subjected to stresses of sufficient magnitude to completely fragment and crush some of the target materials.

The energy involved in comminution is substantial in either solid or loose substrates although considerably less energy is involved in comminuting loose targets. The mass size distribution of comminuted target materials can be described by a simple comminution law (Gault *et al.*, 1963):

$$\frac{m}{Me} = \left(\frac{1}{L}\right)^{a} \tag{3.22}$$

where m is the cumulative mass of fragments smaller than or equal to l, Me

Fig. 3.9. The floor of Tycho may be shock melted rock materials. The framelets forming the photograph are 940 m wide (NASA Photo Lunar Orbiter V-H125 H₂).

is the total ejected mass, L is the size of the largest fragment and a is a constant. This expression is valid over a size range 40 μm < l < L for basalt targets with values of a between 0.3 and 0.6. Below 40 μm a gradually increases. That is, the slope of the cumulative grain-size curve is a function of grain size. Obviously this expression is a first order approximation and could be more realistically approximated by an expression similar to a normal distribution. Epstein (1947) has shown that repetitive fracture of a brittle solid produces a grain-size distribution which asymptotically approaches a log-normal distribution as comminution proceeds. The implications for the development of soil on the lunar surface by multiple hypervelocity impacts is obvious. Braslau (1970) found that the grain-size distribution produced by an impact into loose material (sand) indicated that $a \simeq 2$. Because the detrital target materials were sand grains of a uniform size it is difficult to evaluate Braslau's (1970) results in terms of the Epstein (1947) model. However, it is obvious that the grain size of the sand must be reduced and the standard deviation of the size distribution increased (that is, it becomes more poorly sorted). This aspect of impact is dealt with in Chapter 6 where soil models are covered in more detail.

The effects of comminution on a loose substrate are illustrated graphically by the results of Stöffler $et\ al.$, (1975) in Figure 3.10. The impacted substrate was a loose well-sorted quartz sand with a modal grain size of 3ϕ (125 μm). Following impact the modal class interval was reduced from close to 60 per cent of the mass to less than 40 per cent of the mass in the fall-out ejecta, and less than 20 per cent in the fall-back ejecta which had seen the greatest shock damage. At the same time the shocked sands all became poorly sorted and, in the case of the fall-back ejecta, a second mode

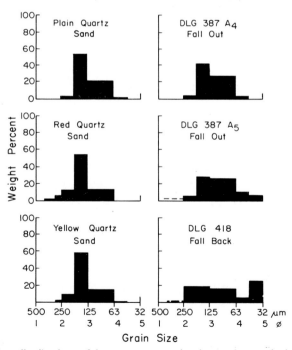

Fig. 3.10. Grain-size distribution of loose quartz sand prior to hypervelocity impact (left) and following impact and ejection (right) (after Stöffler, D., Gault, D. E., Wedekind, J. and Polkowski, G., $Jour.\ Geophys.\ Res.$, 80: 4062-4077, 1975, copyrighted American Geophysical Union).

developed at 5ϕ (32 μm).

    The value of m_b the mass of the largest fragment ejected from a crater has been fairly well established for both large-scale impacts and micro-meteoroid impacts using both natural and experimental craters (Fig. 3.11).

$$m_b = c \, M_e \lambda \tag{3.22a}$$

where c and λ are constants. The value of λ ranges from 1.0 for small craters to 0.8 for large craters. While c ranges from 10^{-1} to 2×10^{-1} for the same conditions. The change in the relationship between m_b and M_e occurs at $M_e > 10^4$ g and is due to the increased probability of secondary fragmentation along pre-existing fractures (joints, etc.) in the country rock and the change of the importance of material strength in larger ejected fragments. That is, it is a scaling effect.

    In Figure 3.12 the grain size distributions of detrital material ejected from 4 microcraters are presented (Gault et $al.$, 1963). The particles cover a wide range of grain sizes (generally poorly sorted) and have particle size distributions similar to ejecta recovered from explosive cratering experiments in desert alluvium (Roberts and Blaylock, 1961). However the importance of

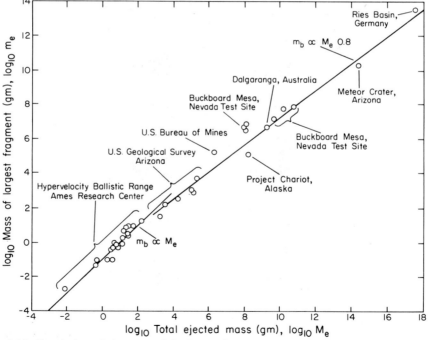

Fig. 3.11. Correlation of the mass of the largest fragment with total ejected mass for explosive and impact cratering events (from Gault et $al.$, 1963).

the strength of materials in relation to scaling is apparent in Figure 3.12 in that the large spalls (L) tend to dominate the coarse end of the grain-size distribution.

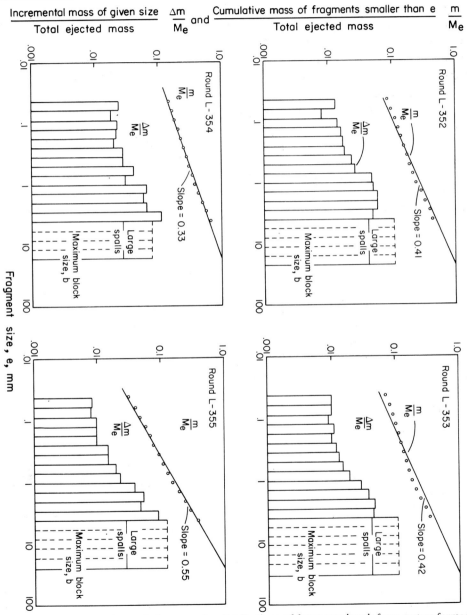

Fig. 3.12. Mass-size and cumulative mass-size distributions of fragmented rock from craters formed in basalt by hypervelocity impact (from Gault *et al.*, 1963).

Ejection and the Development of Stratigraphy

The largest proportion of the kinetic energy of an impacting meteoroid is expended in ejecting material from the crater. The proportion of energy involved also appears to be relatively independent of the nature and strength of the impacted substrate (Table 3.1). The energy imparted to the ejecta is further partitioned in a complex way that determines, to a large extent, the stratigraphic relations of the ejected sediments and their textural properties. Textural properties may be modified either due to abrasion or due to sorting during transport. For convenience ejecta may be separated into three broad categories (1) bulk ejecta, (2) ballistic ejecta and (3) ground flow ejecta. In the past the third category has simply been referred to at the base surge stage, however, as is discussed in more detail later, the occurrence of a base surge is dependent upon a continuing supply of gas without which it will collapse. If the momentum of the flow is sufficient there will be a transition to a grain flow stage with resultant differences in the textural properties of the deposits.

Bulk Ejecta

Near surface strata in the realm of tangential flow at the crater rim are smoothly folded upward and outward with the topmost layers being completely overturned (Gault *et al.*, 1968). All of the strata in the inverted section maintain their identity although the units may be thinned dramatically (Fig. 3.13). Units maintain their identity even in loose non-cohesive substrates and independent of event magnitude (Shoemaker *et al.*, 1963; Gault *et al.*, 1968; Stöffler *et al.*, 1975). In general the bulk ejecta are concentrated close to the crater and at the base of the stratigraphic section (Roberts, 1964).

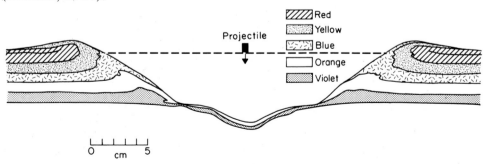

Fig. 3.13. A small experimental crater in a layered cohesionless substrate. The original stratigraphy is preserved as a thin inverted sequence in the ejecta deposit. Compare with figure 3.19 (after Stöffler, D., Gault, D. E., Wedekind, J. and Polkowski, G., *Jour. Geophys. Res.*, *80*: 4062-4077, 1975, copyrighted American Geophysical Union).

Ballistic Ejecta

Lunar gravitational forces are the only forces available to modify the course of an ejected fragment. Consequently an ejected fragment returns to the lunar surface along a segment of an elliptical orbit. In general, with the exception of some limited cases, the perilune of the orbit will be within the body of the moon and the apolune will correspond to the maximum altitude of the ejected fragment above the lunar surface (Gault *et al.*, 1963). The range (R) from the point of impact will then be

$$R = 2r \tan^{-1} \left(\frac{\overline{V}_e{}^2 \sin \theta \cos \theta}{1 - \overline{V}_e{}^2 \cos^2 \theta} \right) \tag{3.23}$$

where $0 \le \theta \le \pi/2$

and $\overline{V}_e{}^2 = V_e{}^2 / rg_0$

V_e is the ejection velocity, g_0 gravitational acceleration (162 cm s^{-2}) at the lunar surface, r the lunar radius (1738 km) and θ the angle of ejection in relation to the local horizontal. The maximum altitude of the ejected particle will then be

$$H_{max} = r \left[\frac{\overline{V}_e{}^2 - 1 + \left\{ 1 - \overline{V}_e{}^2 (2 - \overline{V}_e{}^2) \cos^2 \theta \right\}^{\frac{1}{2}}}{2 - \overline{V}_e{}^2} \right] \tag{3.24}$$

since r and g_0 are known the range and maximum altitude of the ejected particle depends only upon the determination of its vector velocity.

Figure 3.14 shows the development of the ejecta envelope of a small

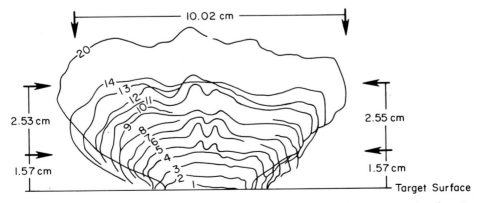

Fig. 3.14. Outline of ejecta envelope traced for first 20 μsec (from Braslau, D., *Jour. Geophys. Res.*, 75: 3987-3999, 1970, copyrighted American Geophysical Union).

experimental impact (6.37 km s^{-1}) for the first 20 μ sec after impact. It can be seen that very little of the detrital material is ejected at angles of less than about 45° (Fig 3.14). Materials ejected at the highest velocities ($\approx$19 km s^{-1}) and lowest angles ($\theta < 20°$) are believed to result from jetting. Jetting is short lived and the ejection velocity quickly decays and the main mass of fragments leaves the crater at velocities less than 0.5 km s^{-1} and with $\theta < 45°$ (Fig. 3.15). The discontinuous variation in the cumulative ejected mass with ejection velocity shown in Figure 3.16 is believed to be associated with a transition from plastic to elastic flow behind the shock front which propagates outward from the point of impact (Gault *et al.*, 1963). The energy involved in the ballistic ejecta seems relatively independent of the properties of the impacted substrate (Braslau, 1970) (Table 3.1). In the absence of an atmosphere the ejected particles can travel large distances around the moon and reach considerable heights above the lunar surface independent of the magnitude of the impact event (Gault *et al.*, 1963) (Figs. 3.17, 3.18).

Most of the ejecta (>90 per cent) will remain below an altitude of 10 km and will return to the lunar surface within a radius of 30 km (Gault *et al.*, 1963). The flux of ejected fragments appears to be 10^3 to 10^5 times that of meteoritic material, although the velocity of the ejecta is two orders of magnitude smaller than that of the impacting projectile. Gault *et al.*, (1963) consequently predicted that the spatial density of projectiles at the lunar surface would be 10^5 to 10^7 times the spatial density of meteoritic projectiles. Horizon glow observations during the Surveyor missions (see Chapter 2) suggest that these predictions are more than reasonable. Gault *et al.* (1963) have also predicted that approximately 1 per cent of the ejected mass escapes the lunar gravitational field and is lost. This is also consistent with the Apollo observations of dust at orbital altitudes around the moon (see Chapter 2).

Possibly the largest unknown factor in lunar sedimentary processes is the role of ballistic ejecta in eroding the lunar surface and producing secondary ejecta. Secondary craters are abundant around the larger lunar craters and as we have seen the ejecta producing them have relatively high velocities. Oberbeck *et al.* (1974) have shown that the ratio M_s of mass ejected from a secondary crater to the mass of the impacting primary crater ejecta that produced any given secondary crater can de determined by

$$M_s = 46.4 \left[\frac{\sin 2\theta}{2 \tan (R_s/3472)} + \sin^2 \theta \right]^{-1} D_s^{-0.401} \cos^{1.134} \theta \qquad (3.25)$$

where the range $R_s = R - R_0/2$, D_s is the measured diameter (km) of a

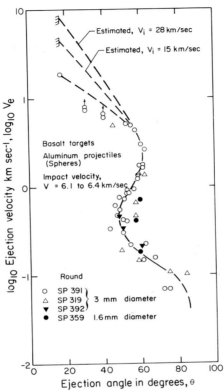

Fig. 3.15. Variation of ejection velocity with ejection angle for fragments thrown out of craters formed in basalt by hypervelocity impact (after Gault *et al.*, 1963).

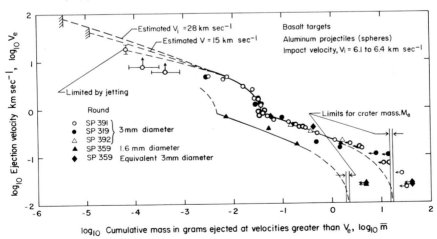

Fig. 3.16. Cumulative mass ejected in excess of a given velocity for fragments thrown out of craters formed in basalt by hypervelocity impact (after Gault *et al.*, 1963).

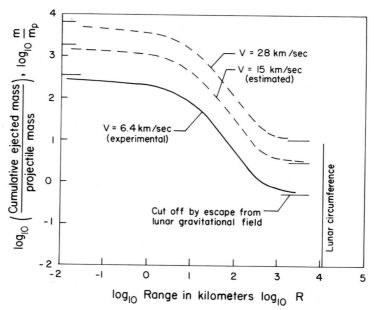

Fig. 3.17. Distribution across the lunar surface of the mass ejected from impact craters formed in basalt (after Gault *et al.*, 1963).

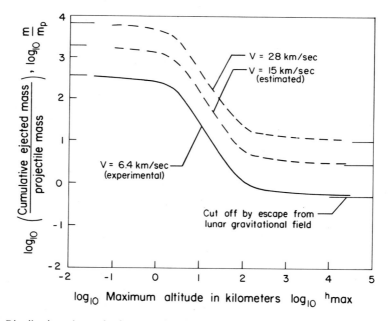

Fig. 3.18. Distribution above the lunar surface of the mass ejected from impact craters formed in basalt (after Gault *et al.*, 1963).

secondary crater at the radial distance R from the center of the primary crater, R_o (km) is the transient crater radius and θ is the ejection angle measured from the local horizontal. For most secondary craters M_s exceeds 1 suggesting that most of these secondary craters produce deposits consisting largely of local material. Secondary cratering may be a major source of detrital materials on the lunar surface.

Stratigraphy and Texture of Ejected Sediments

The earliest insights into the stratigraphy of impact crater ejecta came from the recognition and study of large terrestrial impact craters (Shoemaker, 1960; Bucher, 1963; Short, 1966; Roddy, 1968; Wilshire and Howard, 1968). However, from an experimental view point the most comprehensive study of cratering stratigraphy is that made by Stöffler *et al.* (1975). They impacted 0.3 g projectiles at velocities of 5.9 to 6.9 km s^{-1} into loose quartz sand targets. The results of these experiments provide useful insights into the formation of lunar soil and soil stratigraphy. The experiments involved a target consisting of approximately 1 cm thick layers of colored quartz sand thus allowing a study of the redistribution of individual stratigraphic units about the crater. The impacts produced craters 30 to 33 cm in diameter and approximately 7 to 8 cm in depth. Perhaps surprisingly these small scale experimental craters produced a stratigraphy which is strikingly similar to the larger natural craters such as Meteor Crater, Arizona (Shoemaker, 1960). The bulk ejecta is present forming an overturned flap around the crater rim (Fig. 3.19). This same inverted stratigraphy continues laterally into the ballistic ejecta as well. In general only the upper three sand layers were displaced beyond the ground zero surface in substantial amounts. Some of a fourth layer was displaced within the crater, whereas a fifth layer showed no evidence of having been ejected at all although compaction was considerable. Only the uppermost layer in the target, which corresponds to about 15 per cent of the crater depth, is represented throughout the whole ejecta blanket (Fig. 3.20). Materials from deeper than 28 per cent of the crater depth were all deposited within 2 crater radii. No material from deeper than 33 per cent of the crater depth was ejected beyond the crater rim. The rather conspicuous discrepancy between depth of beyond-rim-ejection and final crater depth is due to about 12.5 per cent compaction in the lower central part of the crater.

A thin layer of material inside the crater preserved the pre-impact stratigraphy of the target. This unit was apparently shock compressed and driven radially downward; it includes shock fused and shock comminuted sand and consequently represents a vertical sequence of zones of decreasing shock metamorphism during impact.

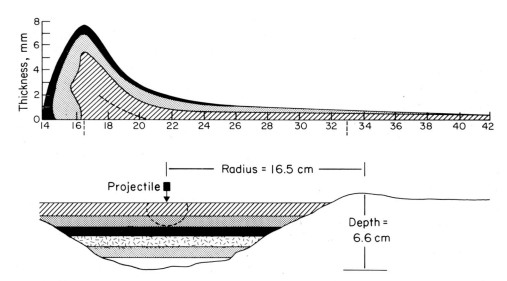

Fig. 3.19. Cross section of crater and ejecta blanket for a horizontally stratified target. Note inverted stratigraphic sequence in the ejecta blanket (after Stöffler, D., Gault, D. E., Wedekind, J. and Polkowski, G., *Jour. Geophys. Res., 80*: 4062-4077, 1975, copyrighted American Geophysical Union).

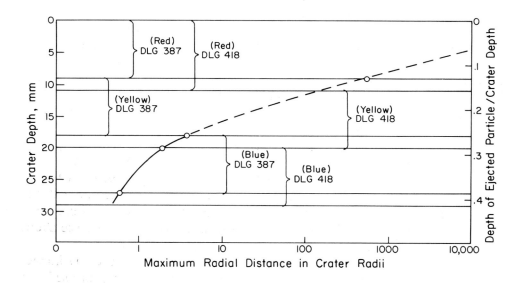

Fig. 3.20. Relationship between pre-impact and post-impact position of particles for a horizontally stratified loose sand target (after Stöffler, D., Gault, D. E., Wedekind, J. and Polkowski, G., *Jour. Geophys. Res., 80*: 4062-4077, 1975, copyrighted American Geophysical Union).

Meteor Crater, Arizona can be compared directly with the small experimental craters produced by Stöffler *et al.* (1975). Rocks exposed at Meteor Crater range in age from the Permian Coconino Sandstone to the Triassic Moenkopi Formation (Figure 3.21). Whereas the pre-impact rocks were horizontally bedded, the beds in the crater walls dip outwards becoming steeper from the bottom of the crater upwards until at the rim they are completely overturned to form the bulk ejecta flap. Rocks now contained in the crater rim were peeled back from the area of the crater much like the petals of a blossoming flower (Shoemaker, 1960). The upturned and overturned strata are intersected by a number of small scissor faults which tend to parallel the regional joint set.

Locally the bulk ejecta may grade outward into disaggregated debris but there is generally a sharp break between the debris and the coherent flap. Outward from the crater the same inverted stratigraphic sequence is present in the debris (Fig. 3.21). The debris consists of poorly-sorted angular fragments ranging from less than a micron to more than 30 m in size. The lowest debris unit consists of Moenkopi Formation and rests directly on the original Moenkopi rocks. This in turn is followed by a unit of Kaibab debris. The contact is sharp close to the crater but at a distance of 830 m from the crater slight mixing occurs along the contact. Finally a unit composed of Coconino and Toroweap Formations rests with sharp contact on the Kaibab debris. Deeper stratigraphic units are not represented in the inverted debris sequence.

Base Surge

The term "base surge" first appears to have been used by Glasstone

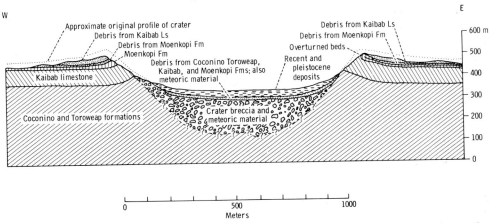

Fig. 3.21. Profile of Meteor Crater, Arizona. Note the overturned flap of Moenkopi Formation on the eastern rim and the inverted stratigraphic sequence preserved in the ejecta blanket (after Shoemaker, 1960).

(1950). The term was applied to a dense ring-shaped cloud which emerged from the base of the cylindrical column of spray ejected vertically by the 1946 nuclear explosion on Bikini Atoll in the south Pacific. Initially the base surge was believed to be due to bulk subsidence of the vertical column ejected by the explosion but after a detailed analysis Young (1965) suggested that the main base surge formed at the crest of a large solitary wave around the edge of the crater formed in the sea by the expanded gases at the explosion center. Later a second smaller base surge developed due to bulk subsidence.

Since the first recognition of the existence of the base surge they have been reported to develop from other large-scale explosions and from explosive volcanic events (Moore *et al.,* 1966; Moore, 1967; Waters and Fisher, 1971; Roberts and Carlson, 1962; Nordyke and Williamson, 1965). Recently there has been a growing body of evidence and opinion which suggests that base surges also develop as a result of hypervelocity impacts and may be one of the most important transportational mechanisms on the lunar surface (O'Keefe and Adams, 1965; Fisher and Waters, 1969; McKay *et al.,* 1970; Lindsay, 1972a, 1972b, 1974). Since the probability of witnessing a large-scale hypervelocity impact is small the following discussion of base surges must, by necessity, be based on observations of experimental explosions and volcanic base surges and then applied to the lunar situation by inference.

Morphology of a Base Surge. A base surge is essentially a density current and behaves in much the same manner as a conventional turbidity current. It is generated at the crater rim following the formation of the rim syncline and the deposition of the bulk ejecta. Expanding gases from the center of the cratering event are then able to move laterally gathering detrital materials as they emerge to form the base surge cloud. The base surge formed during the underwater Bikini test consisted largely of fog and mist. It contained very little calcareous debris from the lagoon bottom. By contrast base surges generated by underwater volcanic events carry large amounts of basaltic glass in addition to water vapor and liquid droplets (Waters and Fisher, 1971). These base surges therefore show considerable differences in behavior. Waters and Fisher (1971) report that on a windless day these volcanic base surges may outwardly appear similar to the Bikini base surge. In a strong wind, however, the jacket of steam hiding the interior of the eruption is blown away. Revealed beneath is a complex core of black glass sand that billows outwards in dark radial plumes, then falls and mixes with the spreading base surge below (Figure 3.22). This implies that in a hypervelocity impact situation much of the ballistic ejecta falls into, and becomes entrained in the expanding gas cloud to form a high density flow

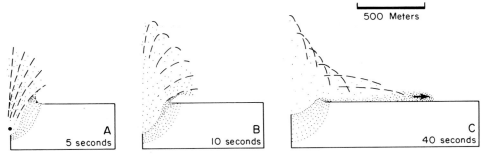

Fig. 3.22. Sequential diagram, with approximate time after burst, showing apparent formation of base surge during underground explosion equivalent to 100 kilotons of chemical explosives. (A) 5 seconds, rock is brecciated around explosion center and expelled generally in a vertical direction forming the vertical explosion column. (B) 10 seconds, crater enlarges and wall surrounding it is overturned allowing outward expansion of gases and ejecta which tear material from overturned flap and base surge. (C) 40 seconds, larger ejecta thrown out by explosion has reached the ground, but base surge continues moving outward as a density flow aided by downslope if present (from Moore, 1967).

lubricated by the entrapped gas. The main difference between a base surge and other density currents is that while a topographic gradient obviously assists the flow its momentum is derived from the kinetic energy of the impact and is independent to a large degree of local topography.

The initial velocity imparted to a base surge can be high. For example the Sedan nuclear explosion which released 4×10^{21} ergs produced a base surge with an initial velocity of 50 m s^{-1}. The distance to which a flow can travel, once initiated, is determined by the initial velocity imparted to it, which is determined by the energy of the event, and the amount of gas trapped within and beneath the flow as a function of its rate of loss. The maximum radial distance (r_b) attained by a base surge is

$$r_b = r_s \, y^{-0.3} \qquad\qquad (3.26)$$

where r_s is the scaled radius of the base surge and y is the yield of the explosion in kilotons (1 KT = 4.2×10^{19} ergs); r_b and r_s are both in meters (Knox and Rohrer, 1963; Rohrer, 1965). The scaled depth of burst (d_s) is determined from

$$d_s = d \, y^{-0.33} \qquad\qquad (3.27)$$

where d is the actual depth of burst and both d and d_s are in meters. r_s can then be determined imperically from Figure 3.23.

There are obvious problems in attempting to relate data from explosions to hypervelocity impacts but it is nevertheless clear that the distance of travel of a base surge is, to a relatively low power, a direct

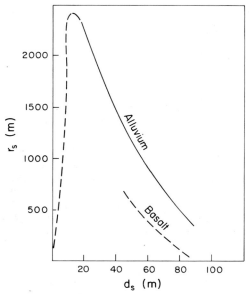

Fig. 3.23. Scaled radius of base surge (r_s) in meters vs. scaled depth of burst (d_s) in meters for explosions in alluvium and basalt (from Rohrer, 1965).

function of kinetic energy. As might be expected the velocity of a base surge declines with time (Fig. 3.24). However, from Figure 3.24 it is apparent that the rate of deceleration is not high. The kinetic energy of the flow is lost through interaction with the surface over which it flows and through particle interaction within the flow. This energy loss is controlled by the rate at which the flow loses its lubricating gaseous medium to the surrounding environment. The volume rate of gas leakage from the bottom of a flow (Q_b) is (Shreve, 1968)

$$Q_b = \frac{kP_b}{m\mu s}\left[1 - \left(\frac{P_t}{P_b}\right)^m\right]\frac{-ks}{\rho K} \tag{3.28}$$

where s is the thickness of the flow, k the harmonic mean permiability of the detrital materials, μ the viscosity of the lubricating gas, P_t and P_b the pressure at the top and bottom of the flow respectively, ρ is the density of the gas at the base of the flow, K is a function of the mass rate of emission per unit bulk volume of detrital material, that is the rate of evolution of gas by the hot particles, and finally m = $(\gamma + 3)/2\gamma$, where γ is the ratio of the heat capacities of the gas at constant pressure and constant volume. In the lunar case P_t is essentially zero and equation 3.28 simplifies to:

$$Q_b = k\left(\frac{P_b}{m\mu s} - \frac{s}{\rho K}\right) \tag{3.29}$$

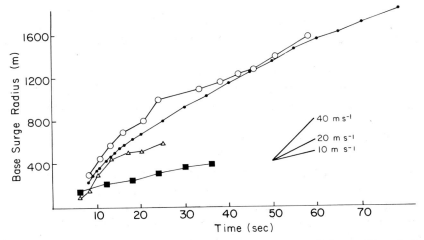

Fig. 3.24. Radial distance of travel of base surge vs. time for 2 test explosions and 2 explosive volcanic eruptions. Circles, Sedan thermonuclear test in alluvium, 100 kilotons; dots, Bikini thermonuclear test in ocean, 20 kilotons; triangles, Myojin Reef eruption of 13:12, September 23, 1952; squares, Anak Krakatau eruption of January 24, 1928 (from Moore, 1967).

If the base surge is travelling on a cushion of gas trapped beneath the flow at the crater rim, Q_b times the travel time gives the loss in thickness of the gas layer due to leakage through the detrital materials. For a trapped layer of lubricating gas to exist the pressure gradient of the leaking gas must be sufficient to support the loose fragments at the base of the flow, that is:

$$\rho_b \, g \leq \frac{\mu \, Q_b}{k_b} \tag{3.30}$$

where ρ_b and k_b are the bulk density and permiability of the flow at its bottom surface and g is gravitational acceleration. Substituting for Q_b the inequality for the lunar case becomes:

$$\frac{k_b \rho_b}{k \rho_a} \leqq \frac{1}{m} - \frac{\mu s^2}{\rho K p_b} \tag{3.31}$$

where ρ_a is the mean bulk density. Provided that m > 1 as would normally be the case the right hand side is always less than 1 regardless of the rate of evolution of gas from hot particles as represented by K (Shreve, 1968). Thus if the permiability and density of the flow were uniform, the pressure gradient in the leaking gas would be too small at the under surface and too large at the upper surface of the flow and simultaneous deposition and fluidization would occur rather than the flow being lubricated by a trapped gas layer. It thus seems unlikely that a base surge travels on a lubricating gas layer for more than a short distance if at all. This is also supported by field evidence, in that base surge deposits begin at or close to the rim of most explosive or volcanic craters.

Fluidization (Leva, 1959) is a physical process by which particles of solid matter are suspended in a fluid by the motion of the fluid. A cushion of gas occurs between the solid particles, so that the mixture behaves like a fluid of relatively low viscosity but of high density (O'Keefe and Adams, 1965). Fluidization occurs in three phases as the flow of gas is increased. First, the *fixed bed phase,* during which the gas flux is small and there is no disturbance of the detrital materials. Second, the *dense phase,* as the gas flux is increased it reaches a critical minimum value called the flow for minimum fluidization at which the character of the detrital materials changes abruptly to a pseudo-fluid. As the flux is increased farther, the flow of gas becomes irregular until finally the velocity of the gas at the surface of the fluidized layer becomes greater than the settling velocity of the particles under Stokes law. Beyond this point the flow regime is in the *dilute phase.* The voidage (ϵ) during the fixed bed phase may be as high as 0.4. At mimimum fluidization the voidage is near 0.6 and at the upper limit of the dense phase the voidage is 0.95 or greater. Voidage (ϵ) relates to density as follows:

$$\rho_1 = (1 - \epsilon) \rho_0 \qquad\qquad (3.32)$$

where ρ_0 is the density of the detrital particles and ρ_1 the bulk density of the fluidized flow.

The pressure gradient through the dense phase of a fluidized flow is (O'Keefe and Adams, 1965)

$$\frac{dp}{dy} = \frac{200 \, v \, \mu \, (1 - \epsilon)^2}{D^2 \, \phi^2 \, \epsilon^3} \qquad\qquad (3.33)$$

where y is the height measured positively upwards, v is the gas velocity, D is the diameter of the detrital particles and ϕ is a shape factor related to sphericity. The strong inverse dependence upon particle size and shape are of significance in the transport of lunar clastic materials. For example, glass particles in lunar soils are frequently extremely irregular in shape resulting in a low value for ϕ thus increasing the absolute value of dp/dy. In contrast the rock and mineral fragments forming the clastic rocks of the Fra Mauro Formation are relatively large and the absolute value of dp/dy is reduced. If the absolute value of dp/dy is large the transition to the dilute phase is much more likely. It is thus more likely that lunar soils are transported in the dilute phase and that there is considerably less particle interaction during transport. The clastic rocks from the Fra Mauro Formation on the other hand show considerable evidence of abrasion in the form of increased roundness due to particle interaction (Lindsay, 1972a, 1972b).

At the base of the flow layer, beneath all of the detrial material, there is a base pressure p_o. For a certain distance above it the pressure will be too low to cause fluidization and the flow will be in the fixed bed phase. Since the gas pressure at the base of the detrital material is continually decreasing with time the fixed bed zone is continually moving upward. That is, sedimentation occurs beneath a fluidized flow at some height (Ymt) which is equivalent to the height at which minimum fluidization occurs. According to O'Keefe and Adams (1965) this height can be determined by assuming that ϵ averages 0.5 over this height and

$$Ymf = \frac{(p_o - p_b)}{0.5\,\rho_o g} \qquad (3.34)$$

P_b can be determined from equation 3.32 by assuming ϵ = 0.6 at minimum fluidization.

Simulation studies show that, under lunar conditions, a voidage (ϵ) of 0.8 is reached at a pressure which is still very high compared to the pressure at the base of the flow (O'Keefe and Adams, 1965). This suggests that the dilute phase is very important in lunar base surges. Because the atmospheric pressure is essentially zero on the moon there will be an explosive increase in the velocity (v) of the escaping gas. Consequently, there will always be a dilute phase at the top of a lunar base surge. Waters and Fisher (1971) describe fine particles and mist drifting downwind from volcanic base surges suggesting that a dilute phase occured above the main body of the flow.

Owing to its greater mobility, and the complete lack of air resistance on the moon, the dilute phase would normally travel faster than the dense phase. At the same time the dilute phase is continually regenerated from the dense phase. That is, any reduction in pressure at the surface of the dense phase would cause the velocity of the gas outflow to increase again and restore the dilute phase pressure. The end result, particularly in the case of small base surges generated in the lunar soil, may be that the whole flow is converted to the dilute phase.

Sorting During Transport. Base-surge deposits and ballistic ejecta are complexly interrelated. In an early phase low-angle ballistic ejecta fall in front of the base surge but later ejecta are either incorporated into the flow or fall onto the base-surge deposits. In all recorded examples of both explosive and volcanically generated base surges the base-surge deposits extend beyond the range of the ballistic ejecta (Fig. 3.25). The nature of a base-surge deposit is determined by processes operating both during transport and at the sediment interface at the time of deposition.

There are three possible mechanisms that can account for the sorting of base surge sediments:

(1) Segregation (that is, sorting in the dense phase). The result of prolonged fluidization is to segregate a flow such that larger particles move downwards (Leva, 1959).

(2) Elutriation (sorting in the transition from the dense to the dilute phase). Elutriation occurs when the velocity of the fluidizing gas exceeds the free-fall velocity of the particles. When presented with a range of particle sizes (impact generated sediments are poorly sorted) the smaller particles will go into the dilute phase before the larger particles if the density is relatively uniform. Elutriation should be most effective in thin base surges and is consequently likely to be most important in the lunar soil. However, elutriation is not highly effective as a sorting mechanism (Leva, 1959). The rate of elutriation is:

$$\log C/C_o = - C_n \, t \tag{3.35}$$

where C_o is the original concentration of fines, C is the final concentration, t is time and C_n is a constant determined by:

$$C_n = \left[(v - v_t)/v_t \right]^{1.3} (D^{0.7}/h_f^{1.4}) \tag{3.36}$$

where D is the particle diameter in inches, h_f is the height of the bed in feet, v the velocity of the gas and v_t the terminal velocity of the particle. For significant elutriation to occur $C_n t$ should be close to 1. The maximum value of the velocity differential in the first parentheses is 1; for $D = 2.5 \times 10^{-3}$ in. (63 μm) and $h_f = 10^2$ ft. (30 m) the elutriation time will be hundreds of hours. Thus elutriation is not likely to cause significant sorting of a dense thick bed.

(3) Sorting in the dilute phase. Though no formulae have been worked out it is clear that there will be a tendency for larger particles to settle out despite the turbulence. Graded beds produced by turbidity currents provide a very useful terrestrial analog (e.g. Kuenen and Migliorini, 1950).

Thus it seems highly probable that sorting occurs in a base surge partly as a result of segregation in the dense phase, partly as a result of elutriation but mostly as a result of differential settling in the dilute phase.

Deposition from a Base Surge. Base surges travel at very high velocities

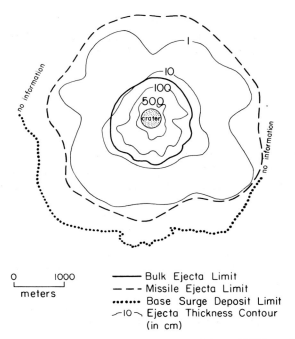

no information

crater

no information

0 1000	———— Bulk Ejecta Limit
|_____|	— — - Missile Ejecta Limit
meters	•••••• Base Surge Deposit Limit
	⌐10⌐ Ejecta Thickness Contour
	(in cm)

Fig. 3.25. Ejecta distribution around Sedan crater (after Nordyke and Williams, 1965 and Carlson and Roberts, 1963).

compared to most other natural fluid flows. In combination with the high velocity the extreme concentration of dense solids, particularly at the base of the flow, results in a high dynamic viscosity and a high shear stress (τ_o) at the sediment interface. Base surges are consequently waning currents travelling under conditions of supercritical flow. A flow is said to be supercritical when its Froud number (F) exceeds unity where:

$$F = \frac{u}{\sqrt{g\,s\,(\rho_1 - \rho_2)/\rho_1}} \qquad (3.37)$$

and u is the velocity of the base surge, s is the thickness of the flow and ρ_1 and ρ_2 are the density of the base surge and any overlying fluid. In the lunar case the density contrast $(\rho_1 - \rho_2)/\rho_1$ goes to unity in the absence of an atmosphere. The Froud number is nondimensional and can be viewed as the ratio between gravity and inertial forces. One result of this relationship is that waves formed on the surface of the fluid are incapable of moving upstream provided the bed remains stable. For the geologist a far more important consequence is that a supercritical flow passing over an erodable bed tends to convert an initial flat bed into a train of smoothly rounded structures known as antidunes (Gilbert, 1914). Antidunes are accompanied by a series of in phase stationary gravity waves on the surface of the fluid (Hand, 1974). Since the Froud number must be at least 1 before antidunes can form it follows from equation 3.37 that at a known velocity u the minimum thickness of the base surge must be:

$$s = \frac{u^2}{g} \frac{\rho_1}{(\rho_1 - \rho_2)} \qquad (3.38)$$

again in the lunar case the density contrast becomes 1 and the thickness depends entirely on velocity and gravity. There is a direct relationship between the amplitude of the antidunes and the surface waves (Hand, 1969).

$$H_b = H_w (1 - 2\pi s/\lambda) \qquad (3.39)$$

where H_b and H_w are the height of the bedforms and flow surface waves respectively, and λ is the wavelength. The wavelength of the antidunes formed by a base surge is directly related to velocity

$$V = \sqrt{(g \lambda/2\pi) (\rho_1 - \rho_2)/(\rho_1 + \rho_2)} \qquad (3.40)$$

again the density term becomes 1 in the lunar case. Field observations of antidunes thus allows direct determination of flow velocity since gravity is the only other variable.

Antidunes are not static sedimentary bodies but may migrate up or down stream in response to differential erosion. Migration results in the development of up- or downstream-dipping cross beds (Fig. 3.26). Alternatively, continuous laminae that are draped over the whole dune or plane bedforms that are characteristic of the upper phase may develop depending upon conditions at the sediment interface. The relationship between bedforms formed during supercritical and subcritical flow is shown in Figure 3.27, which is one of a large number of two-dimensional predictive schemes (Allen, 1968). The diagram implies that a waning current should produce a regular heirarchy of bedforms; antidunes $\rightarrow$ plane bed $\rightarrow$ large scale ripples and possibly small scale ripples before coming to rest. As will be seen in the following section, there are observational data to support this view, although base surges generally produce antidunes and plane beds. Further, in reality bedforms must ultimately be treated as a multidimensional problem, a two dimensional treatment is at best only a first approximation (Southard, 1971; Hand, 1974). Small ripples can form on the surface of antidunes during supercritical flow, an indication that antidune formation (supercritical flow over an erodable bed) is in reality, to some extent, independent

Flow Direction

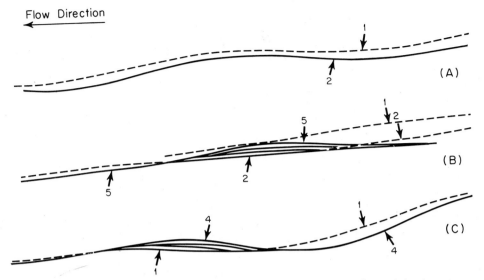

Fig. 3.26. Bed profiles showing evolution of antidunes generated by experimental density currents. Dashed portions were subsequently eroded. Numbers indicate time sequence. In (a), the antidune was produced by differential erosion; in (b), by accretion; in (c), by a combination of erosion and accretion (after Hand, 1974).

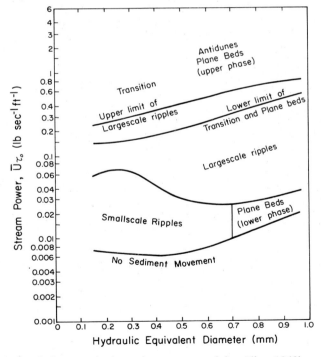

Fig. 3.27. Bedforms in relation to grain size and stream power (after Allen, 1968).

of the parameters that control ripples (transport characteristics of bed material, together with velocity conditions near the bed) (Hand, 1974). It also seems highly unlikely that a base surge would travel more than a very short distance in subcritical flow simply because of its dependence upon the entrapped gas for lubrication.

Dune-like structures that form concentric patterns around many lunar craters are perhaps the most convincing evidence of the occurence of base surge on the moon (Figure 3.28). For example, the dune structures around the crater Mösting C have a wavelength of approximately 400 m at a distance of 1 km from the crater rim. From equation (3.40) it follows that if the dunes are base surge antidunes, as seems reasonable, the base surge was travelling at approximately 10 m s^{-1} or slightly more than 36 km hr^{-1}. The minimum flow thickness would then be approximately 64 m. The figures are clearly reasonable in terms of terrestrial experience.

Base Surge Sediments. Clastic rocks which are unquestionably of base surge origin have not yet been studied on the lunar surface. Consequently, it is important to look at terrestrial deposits of known base surge origin as a basis for interpretations presented in following chapters. As a result

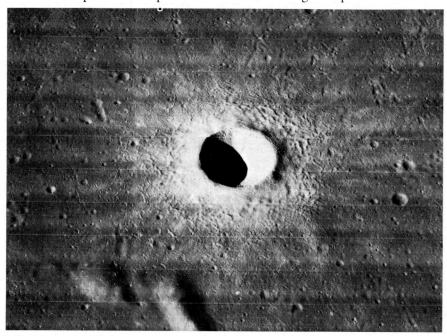

Fig. 3.28. The crater Mösting C which is 3.8 km in diameter has numerous concentrically disposed dune structures on its ejecta blanket. These dunes offer clear evidence of the role of base surge in the transportation and deposition of lunar sediments. Blocky ballistic ejecta are also visible on the crater rim (NASA Photo Lunar Orbiter III-M113).

of the fact that base surges have only been so recently recognized the amount of literature on their deposits is limited. Moore (1967) summarized much of the literature to that date on base surges generated both by volcanic eruptions and nuclear explosions. Crowe and Fisher (1973) have also provided useful summary data concerning textural properties of known base surge deposits and described some older deposits of probable base surge origin in Death Valley, California.

Base surge deposits are generally characterized by one or more of the following megascopic features (Crowe and Fisher, 1973):

(1) Low-angle cross-bedding, which is particularly diagnostic if the flows can be shown to have moved radially from the vent or explosion center, or if there is evidence of upslope movement.

(2) The deposits are generally lenticular. On a large scale the thickness of the deposit is in the first place determined by topography. This is particularly significant where deposits can be seen to thicken markedly in short distances from topographic highs into lows. On a small scale, beds pinch or swell and split and coalesce when traced laterally.

(3) Ejecta layers are plastered against topographic features at angles greater than the angle of repose. These beds are frequently thinly bedded and indicate the cohesive nature of the debris.

(4) Thin and relatively continuous bedding occurs in deposits around the vent.

(5) Finally, large-scale antidune structures may be preserved intact. Antidunes should show a regular radial pattern of decreasing wavelength (due to decreasing velocity) away from the source of the base surge.

Grain Size of Base Surge Deposits. Most base surge sediments are poorly sorted. However, as in most density current deposits, there are consistencies in the grain size parameters relating to the transportational mechanism and conditions at the depositional interface. On the large scale it can be seen from table 3.2 that there is a relationship between distance of travel (and hence velocity and antidune wavelength) and grain size of sediments deposited in antidune foresets and backsets. In general grain size decreases with distance of travel and the decreasing velocity of the base surge. Coincident with the decreasing grain size is an increase in sorting and a general thinning of the deposits.

On the small scale grain-size parameters are also related to the nature of the bed and the current structures into which they are incorporated. From Figure 3.29 it can be seen that grain-size parameters separate plane beds from antidunes, indicating that antidune deposits are finer grained. From Table 3.2 it is apparent that forset beds are finer grained and better sorted than backset beds from the same antidune. Crowe and Fisher (1973)

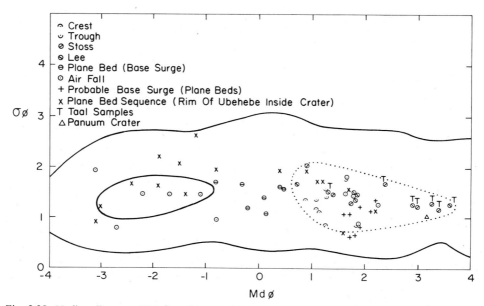

Fig. 3.29. Median diameter (Md ϕ) and Inman (1952) sorting ($\sigma\phi$) plots of base surge deposits. Seven Taal Volcano (Philippines) samples and one sample from cross-bedded base-surge deposits from Panuum Crater near Mono Lake, California (after Crowe and Fisher, 1973).

TABLE 3.2

Grain-size parameters of sediments deposited in antidunes during the 1965 Taal Volcano eruption in the Phillipines.

Sample	Median Diameter Md ϕ	Standard Deviation $\delta\ \phi$	Distance from Vent km	Antidune Wavelength m	Calculated Velocity m s^{-1}	Thickness of bedding set m	Total Thickness of Deposit
Backset	0.9	2.01	0.75	11.5	4.24	2.0	4.0
Foreset	1.3	1.64					
Backset	2.4	1.53	1.0	9.0	3.75	1.0	2.5
Foreset	2.9	1.24					
Backset	3.3	1.19	1.5	5.5	2.93	0.7	1.0
Foreset	3.0	1.20					
Backset	3.4	1.16	2.0	4.0	2.50	0.5	0.5
Foreset	3.6	1.21					

experienced difficulty in separating air fall deposits from plane-bed base surge deposits. While air fall deposits do not exist on the moon it should perhaps be kept in mind that the dilute phase of the base surge must be very important in the lunar case and may result in finer-grained deposits, similar to air-fall deposits on the earth.

Sedimentary Structures. Possibly the most complete description of the sedimentary structures in base-surge deposits is that given by Crowe and Fisher (1973) for the Ubehebe Craters in California. At this locality inclined laminations were found to occur within single beds which have planar upper and lower surfaces or within dune- or ripple-like structures which have undulating upper surfaces and planar to gently-curved lower contacts. Similar undulating structures have also been described from base surge deposits at Taal Volcano, Philippines (Moore, 1970; Waters and Fisher, 1971).

Single dune structures begin their development on plane parallel surfaces with a slight heaping up of debris with low-angle lee-side laminae and without stoss-side laminae. These small ripple-like features are followed by distinctly finer-grained tuffs showing nearly symmetrical low-angle lee-side and stoss-side laminae. The crests in successive laminae usually migrate downcurrent, although in some dune structures one or two of the lower laminae tend to steepen. This steepening is usually accomplished by an upward increase in the angle of climb, although the angle may be variable within single dune structures. The set of laminae on the stoss side tends to be thinner and finer grained than the lee-side set. Therefore, erosion of the stoss side must have occured during migration and only the finer grained and the most cohesive sediment collected on that side. The lee-side set includes coarse-grained tuff and lapilli tuff units which alternate with fine-grained laminae similar in grain size to laminae on the stoss side; the fine-grained layers are commonly continuous from stoss side to lee side. Apparently the coarser ash was less cohesive than the fine ash and was more easily swept from the stoss-side surface onto the lee side and was preserved there.

Although the stream gradients are relatively constant within any one segment, individual beds or laminae on which the small dunes are built are somewhat undulatory. The slope angles of the depositional surfaces underlying the dunes therefore vary within a few degrees, but commonly dip downstream, although they may be horizontal or dip upstream. Plots of wave-length to wave-height ratios versus the angle of depositional surfaces on which the structures are built (Fig. 3.30) show that wave length tends to increase with respect to height as the uphill depositional slope increases. Plots of stoss-side versus lee-side dips of the small dunes for which depositional slope is not compensated (Fig. 3.31a) show that most of the

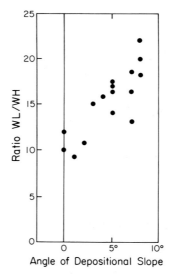

Fig. 3.30. Plot of the ratio wave length/wave height versus angle of depositional slope for small dune-like structures southeast of Ubehebe Crater (after Crowe and Fisher, 1973).

stoss-side dips are greater than the lee-side dips. The reverse is true if slope is calculated at 0° (Fig. 3.31b).

The dimensions of the antidunes are extremely variable. Dune-like structures with wave lengths from 30 to 210 cm and wave heights from 2.5 to 23.8 cm occur at Ubehebe craters (Crowe and Fisher, 1973). At Taal Volcano there are dunes with wave lengths of up to 11.5 m (Waters and Fisher, 1971). Clearly these structures are all closely related despite the range in size because wavelength and amplitude appear to be correlated (Fig. 3.32). It is evident from equations 3.39 and 3.40 that flow velocity and thickness are the main controlling variables although other variables such as grain size and depth of flow must also be involved.

The Ubehebe dunes bear some similarities to ripple-drift cross laminations in that they show a migration of successive wave forms up the stoss side of preceding wave forms, down current crest migration, varying degrees of thickness of the stoss-side lamination set and upward increase in climb angle (Fisher and Crowe, 1973). The upward sequence of cross lamination observed in some large dune structures is mindful of Allen's (1970) predicted appearance of ripple-drift types in unsteady flow. It should however be kept in mind that the Ubehebe bedforms all have much larger wave heights and wave lengths than typical ripple-drift cross-lamination.

Deposition at a particular point should reflect the energy loss of the base surge with time. Thus single base-surge bedding sets are composite structures reflecting a sequential upward change in flow conditions. The

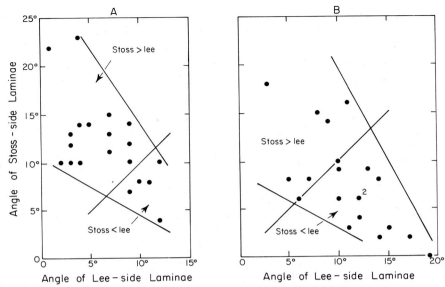

Fig. 3.31. Angle of stoss-side laminae plotted against angle of lee-side laminae for small dune-like structures southeast of Ubehebe Crater. (A) Plot without correction for depositional slope; (B) plot with correction for depositional slope. Lines sloping toward left on diagrams A and B are where stoss-side dips equal lee-side dips; other lines bound areas of plotted points (from Crowe and Fisher, 1973).

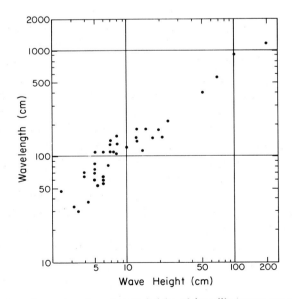

Fig. 3.32. Log-log plot of wave length and wave height of dune-like structures measured at Ubehebe and of four examples from Taal Volcano, Philippines (after Crowe and Fisher, 1973).

Ubehebe dunes described by Crowe and Fisher (1973) preserve a sequence of high-flow-regime bed forms which appear to have been preserved because of rapid deposition and burial and high sediment cohesion. The bedforms appear to progress through the antidune phase to sinuous laminations and plane bedforms as stream power decreased.

Grain Flow

The base surge model fails to account for two, possibly partially dependent, variables in the depositional process: (1) scaling effects (2) the supply of lubricating gas. The smallest event described in the literature as producing a base surge is a 220 kg chemical explosion mentioned by Moore (1967) although there is undoubtably much more literature on the subject which is unavailable. There is no record of an experimental hypervelocity impact producing anything similar to a base surge. Somewhere between the two extremes is a transitional area determined by the scale of the event in relation to the grain size of the ejecta (and thus permiability of the ejecta mass) and the velocity and viscosity of the escaping gas. In smaller events the energy associated with the expanding gas probably contributes to essentially "blowing" the detrital materials away but the escaping gases are not compressed or trapped by the detrital materials long enough to produce a density flow.

The second variable relevent to this discussion is the gas supply during the flow. The gas content of a base surge is not necessarily a direct function of the momentum imparted to the base surge during crater excavation. To a large extent, under a single set of conditions, the volume of gas generated will increase as the kinetic energy of the impact event increases but the actual volume of gas generated and then trapped within the base surge depends largely on the nature of the bedrock. For example, experimental explosions in alluvium produce base surges which travel approximately twice as far as base surges from explosions in basalt, largely because of the higher moisture content and the smaller grain size of the alluvium (Moore, 1967).

The effect of a dispersion of solid detrital particles on the shear resistance of a fluid was first investigated by Einstein (1906) and later by Jeffreys (1922) who also studied the motion of the particles suspended in the fluid. More recently Bagnold (1954) has extended those studies and applied the results to sediment transport.

The linear concentration (Li) of particles in a fluid can be expressed in terms of the voidage (ϵ) as follows:

$$Li = \frac{1}{\dfrac{(1 - \epsilon_o)^{1/3}}{(1 - \epsilon)} - 1} \qquad (3.41)$$

where ϵ_o is the minimum voidage for close packed spheres. The chances of shear stress being transmitted by continuous and simultaneous contact between many grains should thus decrease rapidly as Li decreases, that is, as the voidage ϵ or the gas content of the flow increases. Bagnold (1954) has shown that when Li is large and the effects of grain inertia dominate a dispersive pressure P is generated in the flowing mass of detrital material such that

$$P = a_i \rho_b \, \text{Li} f \, (\text{Li}) \, D^2 \, (du/dy)^2 \, \text{Cos} \, a_i \qquad (3.42)$$

Along with P a proportional grain shear stress T of magnitude

$$T = P \tan a_i \qquad (3.43)$$

is generated where a_i is some unknown angle determined by the collision conditions. D and ρ_b are particle diameter and density respectively, du/dy is shear strain and y is measured positively above the sediment interface. The grain shear stress T is additional to any residual stress τ' due to momentum transfer within the intergranular fluid. Equation 3.42 gives the dispersive grain pressure normal to the direction of shear. It should be noted that the density of the entrapped fluid ρ does not enter into the equation. When the effects of grain inertia dominate (Li is large, possibly >12) the shear stress T is purely a grain stress to be regarded as additive to any fluid shear stress τ' Consequently the mean shear stress $T = T + \tau'$. However, when fluid viscosity dominates where Li is small (<12) a mixed shear stress due to the effects of fluid viscosity modified by the dispersed grains occurs and

$$T = \mu \, (1 + \text{Li}) \, (1 + f'(\lambda)/2) \, du/dy \qquad (3.44)$$

From equations 3.34 and 3.35 from the previous discussion of base surge it can be seen that the pressure changes vertically and that the sediment interface moves upward with time. If the gas supply is limited the base of the flow will enter the grain-inertia region while the momentum of the flow is still large. Consequently du/dy and Li will both be large. This suggests that small scale hypervelocity impact events and events occuring in gas-poor bedrock materials could produce density flows which either travel entirely in the grain-inertia region or rapidly pass into that region.

In the inertia region of grain flow the dispersive pressure P is a function of D^2 at a given shear stress du/dy. This suggests that in less-well-sorted sediments the larger grains should tend to drift towards the zone of least shear strain (i.e. towards the free surface of the flow) and the smaller grains should move towards the bed where the shear strain is greatest. The result is that small scale hypervelocity impact events and gas-poor base surges should produce reverse graded units. There is little experimental data to support this but Lindsay (1974) has found evidence to suggest that grain flow may be important in the deposition of thin (< 3 cm) units found in the lunar soil. This aspect of grain flow is discussed in more detail in Chapter 6.

Thickness of Ejecta Blankets

The ejecta blanket surrounding a crater consists of material deposited by three different but interrelated mechanisms. All three mechanisms produce deposits which thin away from the crater rim. The radial variation in thickness (t_b) of an ejecta blanket can be described by a simple power function of the range expressed in a dimensionless form (r/R) where R is the radius of the crater and r the distance measured radially from the center of the crater (McGetchin *et al.*, 1973).

$$t_b = T_o \, (r/R)^{-\beta} \tag{3.45}$$

where T_o is the ejecta thickness at the rim crest and β is a constant. T_o can be estimated using the earlier function for rim height (R_f) developed by Baldwin (1963), or from a similar expression developed from experimental explosion craters and terrestrial and lunar impact craters by McGetchin *et al.* (1973).

$$T_o = 0.14 \, R^{0.74} \tag{3.46}$$

The evaluation of β appears to be more complex and values from 2.0 to 3.7 have been calculated. However, calculated estimates using data from large lunar craters suggest that β is close to 3.0. Combining equations 3.45 and 3.46 and integrating yeilds a total volume V_T of:

$$V_T = 2\pi T_o R^2 \tag{3.47}$$

The above equations provide a means for evaluating lunar stratigraphic relations. The thickness estimates are average values assuming radial symmetry; variables such as antecedent topography are not accounted for.

References

Allen, J. R. L., 1968. *Current Ripples*. North Holland Pub. Co., 433 pp.

Allen, J. R. L., 1970. A quantitative model of climbing ripples and their cross laminated deposits. *Sedimentology, 14:* 5-26.

Bagnold, R. A., 1954. Experiments on a gravity free dispersion of large solid spheres in a Newtonian fluid under shear. *Proc. Royal Soc. Sec. A., 225:* 49-63.

Baldwin, R. B., 1949. *The Face of the Moon*. Univ. Chicago Press, Chicago, 239 pp.

Baldwin, R. B., 1963. *The Measure of the Moon*. Univ. of Chicago Press, Chicago, 488 pp.

Beals, C. S., 1965. The identification of ancient craters. *Ann. N. Y. Acad. Sci., 123:* 904-914.

Braslau, D., 1970. Partitioning of energy in hypervelocity impact against loose sand targets. *Jour. Geophys. Res., 75:* 3987-3999.

Bucher, W. H., 1963. Cryptoexplosion structures caused from without or within the earth ("Astroblems" or "geoblemes"?). *Am. Jour. Sci., 261:* 597-649.

Charters, A. C., 1960. High speed impact. *Sci. Am., 203:* 128-140.

Crowe, B. M. and Fisher, R. V., 1973. Sedimentary structures in base surge deposits with special reference to cross-bedding, Ubehebe Craters, Death Valley, California. *Geol. Soc. Amer. Bull., 84:* 663-682.

Dence, M. R., 1964. A comparative structural and petrographic study of probable Canadian meteorite craters. *Meteoritics, 2:* 249-270.

Dence, M. R., 1965. The extraterrestrial origin of Canadian craters. *Ann. N. Y. Acad. Sci., 123:* 941-969.

Dence, M. R.. 1968. Shock zoning at Canadian craters: Petrography and structural implications. In: B. M. French and N. M. Short, Mono, Baltimore (Editors), *Shock Metamorphism of Natural Materials,* 169-184.

Dence, M. R., 1971. Impact melts. *Jour. Geophys. Res., 76:* 5552-5565.

Einstein, A., 1906. Eine neue Bestimmung der Molekuldimensionen. *Ann. Phys., Ser. 4, 19:* 289-306.

Engelhardt, W. von, 1972. Impact structures in Europe. *Internat. Geol. Cong., Sect. 15 (Planetology),* Montreal, Canada, 90-111.

Epstein, B., 1947. The mathematical description of certain breakage mechanisms leading to the logarithmic−normal distribution. *Jour. Franklin Inst., 44:* 471-477.

Fisher, R. V., 1966. Mechanisms of deposition from pyroclastic flows. *Amer. Jour. Sci., 264:* 350-363.

Fisher, R. V. and Waters, A. C., 1969. Bed forms in base-surge deposits: lunar implications. *Science, 165:* 1349-1352.

Fisher, R. V. and Waters, A. C., 1970. Base surge bed forms in Maar Volcanoes. *Amer. Jour. Sci., 268:* 157-180.

French, B. M., 1970. Possible relations between meteorite impact and igneous petrogenesis, as indicated by the Sudbury structure, Ontario, Canada. *Bull. Volcanol., XXXIV-2:* 466-517.

Gault, D. E., 1973. Displaced mass, depth, diameter, and effects of oblique trajectories for impact craters formed in dense crystalline rocks. *The Moon, 6:* 32-44.

Gault, D. E. and Heitowit, E. D., 1963. The partitioning of energy for hypervelocity impact craters formed in rock. *Proc. Sixth Hypervelocity Impact Symp.,* Cleveland, Ohio, *2:* 419-456.

Gault, D. E., Hörz, F., and Hartung, J. B., 1972. Effects of microcratering on the lunar surface. *Proc. Third Lunar Sci. Conf., Suppl. 3, Geochim. Cosmochim. Acta, 3:* 2713-2734.

Gault, D. E., Quaide, W. L. and Oberbeck, V. R., 1968. Impact cratering mechanics and structures. In: B. M. French and N. M. Short, Mono, Baltimore (Editors), *Shock Metamorphism of Natural Materials,* 87-100.

Gault, D. E., Shoemaker, E. M., and Moore, H. J., 1963. Spray ejected rrom the lunar surface by meteoroid impact. NASA TN D-1767, 39 pp.

Gault, D. E., Hörz, F., Brownlee, D. E. and Hartung, J. B., 1974. Mixing of lunar regolith. *Proc. Fifth Lunar Sci. Conf., Suppl.5, Geochim. Cosmochim. Acta, 3:* 2365-2386.

Gilbert, G. K., 1893. The moon's face, a study of the origin of it features. *Philos. Soc. Washington Bull., 12:* 241-292.

Gilbert, G. K., 1914. Transportation of debris by running water. *U. S., Geol. Survey Prof. Paper 86,* 263 pp.

Glasstone, S. ed., 1950. The effects of atomic weapons. *Los Alamos, N. M., Los Alamos Sci. Lab., U. S. Atomic Energy Comm.,* 465 pp.

Green, J., 1965. Hookes and Spurrs in selenology. *Ann. N. Y. Acad. Sci., 123:* 373-402.

Green, J., 1965. Tidal and gravity effects intensifying lunar defluidization and volcanism. *Ann. N. Y. Acad. Sci., 123:* 403-469.

Green, J., 1970. Significant morphologic features of the moon. *Image Dynamics, 5:* 29-30.

Green, J., 1971. Copernicus as a lunar caldera. *Jour. Geophys. Res. 76:* 5719-5731.

Hand, B. M., 1974. Supercritical flow in density currents. *Jour. Sediment. Pet., 44:* 637-648.

Head, J. W., 1974. Orientale multi-ringed basin interior and implications for the petrogenesis of lunar highland samples. *The Moon, 11:* 327-356.

Jeffreys, G. B., 1922. The motion of ellipsoidal particles immersed in a viscous fluid. *Royal Soc. (Lon.) Proc., Ser. A, 102:* 161-179.

Knox, J. B. and Rohrer, R., 1963. Project Pre-Buggy base surge analysis. *Univ. Calif. Lawrence Radiat. Lab.* PNE-304, 37 pp.

Kuenen, Ph. H. and Migliorini, G. I, 1950. Turbidity currents as a cause of graded bedding. *Jour. Geol., 58:* 91-126.

Leva, M., 1959. *Fluidization.* McGraw-Hill, New York, 327 pp.

Lindsay, J. F., 1972a. Sedimentology of clastic rocks from the Fra Mauro region of the moon. *Jour. Sediment. Pet., 42:* 19-32.

Lindsay, J. F., 1972b. Sedimentology of clastic rocks returned from the moon by Apollo 15. *Geol. Soc. Amer. Bull., 83:* 2957-2870.

Lindsay, J. F., 1974. Transportation of detrital materials on the lunar surface: Evidence from Apollo 15. *Sedimentology, 21:* 323-328.

Lindsay, J. F. and Srnka, L. J., 1975. Galactic dust lanes and the lunar soil. *Nature, 257:* 776-778.

Mandeville, J. C. and Vedder, J. F., 1971. Microcraters formed in glass by low density projectiles. *Earth Planet. Sci. Lett., 11:* 297-306.

McGetchin, T. R., Settle, M. and Head, J. W., 1973. Radial thickness variation in impact crater ejecta: Implications for lunar basin deposits. *Earth Planet. Sci. Letters, 20:* 226-236.

McKay, D. S., Greenwood, W. R., and Morrison, D. A., 1970. Origin of small lunar particles and breccia from the Apollo 11 site. *Proc. Apollo 11 Lunar Sci. Conf., Suppl. 1, Geochim. Cosmochim. Acta, 1:* 673-693.

Moore, H. J., Gault, D. E. and Heitowit, E. O., 1965. Change in effective target strength with increasing size of hypervelocity impact craters. *Proc. Seventh Hypervelocity Impact Symp., 4:* 35-45.

Moore, H. J., Gault, D. E. and Lugn, R. V., 1961. Experimental hypervelocity impact craters in rock. *Proc. Fifth Hypervelocity Impact Symp., 1:* 625-643.

Moore, J. G., 1967. Base surge in recent volcanic eruptions. *Bull. Volcanol., 30:* 337-363.

Moore, J. G., Nakamura, K. and Alcaraz, A., 1966. The 1965 eruption of Taal Volcano. *Science, 151:* 955-960.

Nakamura, Y., Latham, G., Lammlein, D., Ewing, M., Duennebier, F. and Dorman, J., 1974. Deep lunar interior inferred from recent seismic data: *Geophys. Res. Letters, 1:* 137-140.

Nordyke, M. D. and Williamson, M. M., 1965. The Sedan event. *Univ. Calif. Lawrence Radiat. Lab.,* PNE-242F, 114 pp.

Oberbeck, V. R., Morrison, R. H., Hörz, F., Quaide, W. L. and Gault, D. E., 1974. Smooth plains and continuous deposits of craters and basins. *Proc. Fifth Lunar Sci. Conf., Suppl. 5, Geochim. Cosmochim. Acta, 1:* 111-136.

O'Keefe, J. A. and Adams, E. W., 1965. Tektite structure and lunar ash flows. *Jour. Geophys. Res., 70:* 3819-3829.

Pike, R. J., 1974. Craters on the earth, moon and Mars: Multivariate classification and mode of origin. *Earth Planet. Sci. Lett., 22:* 245-255.

Roberts, W. A., 1964. Notes on the importances of shock craters lips to lunar exploration. *Icarus, 3:* 342-347.

Roberts, W. A. and Blaylock, J. A., 1961. Distribution of debris ejected by the Stagecoach series of high explosive cratering bursts. *Boeing Airplane Co., Document D2-6955-1.*

Roberts, W. A. and Carlson, R. H., 1962. Ejecta studies, Project Sedan. *Univ. Calif. Lawrence Radiat. Lab., PNE-217P,* 62 pp.

Roddy, D. J., 1968. The Flynn Creek Crater, Tennessee. In: French, B. M. and Short, N. M. (Editors), *Shock Metamorphoism of Natural Materials,* Mono, Baltimore, 291-322.

Rohrer, R., 1965. Base surge and cloud formation—Project pre-Schooner. *Univ. Calif. Lawrence Radiat. Lab., PNE-503F,* 10 pp.

Shoemaker, E. M., 1960. Penetration mechanics of high velocity meteorites, illustrated by Meteor Crater, Arizona. *Internat. Geol. Cong. XXI Session, 18:* 418-434.

Shoemaker, E. M., 1962. Interpretation of lunar craters. In: Z. Kopal (Editor), *Physics and Astronomy of the Moon,* Academic Press, N. Y., 283-360.

Short, N. M., 1965. A comparision of features characteristic of nuclear explosion craters and astroblemes. *Ann. N. Y. Acad. Sci., 123:* 573-616.

Short, N. M., 1966. Shock-lithification of unconsolidated rock materials. *Science, 154:* 382-384.

Shreve, R. L., 1968. Leakage and fluidization in air-layer lubricated avalanches. *Geol. Soc. Amer. Bull., 79:* 653-658.

Southard, J. B., 1971. Presentation of bed configurations in depth-velocity-size diagrams. *Jour. Sediment.Pet., 41:* 903-915.

Spurr, J. E., 1944-1949. *Geology Applied to Selenology.* Science Press, Lancaster (4 volumes).

Stöffler, D., Gault, D. E., Wedekind, J. and Polkowski, G., 1975. Experimental hypervelocity impact into quartz sand: Distribution and shock metamorphism of ejecta. *Jour. Geophys. Res., 80:* 4062-4077.

Urey, H. C., 1952. *The Planets.* Yale Univ. Press, 245 pp.

Vedder, J. F., 1972. Craters formed in mineral dust by hypervelocity micro-particles. *Jour. Geophys. Res., 77:* 4304-4309.

Waters, A. C. and Fisher, R. V., 1971. Base surges and their deposits: Capelinhos and Taal Volcanoes. *Jour. Geophys. Res., 76:* 5596-5614.

Wilshire, H. G. and Howard, K. A., 1968. Structural patterns in central uplifts of cryptoexplosive structures as typified by Sierra Madera. *Science, 162:* 258-261.

Young, G. A., 1965. The physics of base surge. *U. S. Naval Ordinance Lab.,* Silver Spring, Md., 284 pp.

CHAPTER 4

Stratigraphy and Chronology of the Moon's Crust

Lunar Topography

Even when viewed from earth, the moon's surface is easily divided into mare and terra (or highland) areas on the basis of albedo. Upon closer scrutiny we find that highland areas are considerably rougher and more intensely cratered than the mare suggesting a considerable difference in age. This two-fold division of the lunar surface, first suggested by Galileo in 1610, is very basic in the understanding of the nature of the moon.

Laser altimetry data collected from orbit by the Apollo 15, 16 and 17 spacecraft has shown that the mare areas are, for the most part, topographic lows although not all mare surfaces are at the same altitude (Kaula, 1972, 1973, 1974) (Fig. 4.1). For example, Mare Crisium and Mare Smythii are extremely low (-5 km) whereas Oceanus Procellarum is not much lower than the surrounding highlands (-3 km vs -2 km). Overall the mare are 3—4 km lower than the terra. In Figure 4.2 a hypsometric curve for the moon based on the altimeter data is compared with the earth. A direct comparison of the two curves is difficult because the lunar curve is based on topographic deviations about a mean sphere centered at the center of the lunar mass. This center of mass is not coincident with the geometric center of the moon but is displaced 2—3 km towards the lunar near side at about 25°E. Terrestrial data are related to a complex mean geoid (i.e. mean sea level) rather than a simple sphere. However, the curves are grossly similar, except that the lunar curve is flatter, indicating that the topographic distinction between highland and mare areas is less clear than the division between ocean and continent on earth. It is tempting to suggest that the similarity of the two curves indicates that something similar to ocean floor spreading occurred during the early formation of the lunar crust but it is probably more realistic to suggest that the shape of the lunar curve resulted from the excavation of the mare by impact and a resultant redistribution of the crustal materials.

As a result of the extensive reworking of the lunar surface the moon's crust is structurally and stratigraphically complex and much of the early record of crustal evolution has been obliterated.

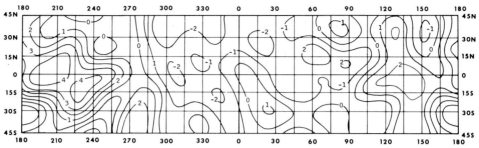

Fig. 4.1. Laser altimetry of the lunar surface. The mare are generally lower than the terra.

Early History of the Moon

The moon may be as old as ≈4.65 aeons (Tatsumoto, 1970; Nunes *et al.*, 1973). This age has been derived from U-Th-Pb systematics of lunar soils and soil breccias. However, while soils and soil breccias offer representative samples of large areas of the lunar crust, they can not be regarded as representative of the moon as a whole. Tera and Wasserburg (1974), while not suggesting that 4.65 aeons is an unreasonable figure for the age of the moon, suggest that the data are too limited at present to be certain. The approximate age of the solar system and the planets is however, reasonably well defined at ≈4.5 aeons with an uncertainty of ≈200 m.y. in the precise age of the planets. Presumably the moon originated within this time period.

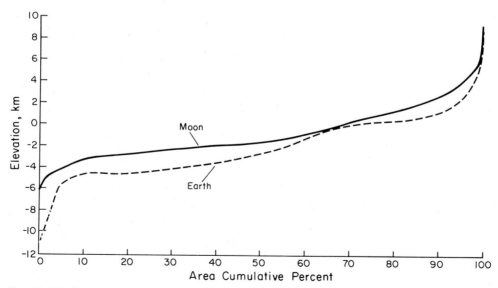

Fig. 4.2. Distribution of the surface of the moon and earth in terms of percentage of surface below a given altitude. Note the similarity between the two curves and the fact that the lunar curve is flatter (Lunar data courtesy of L. Srnka and W. Wollenhaupt).

There is little radiometric data concerning the early events in the formation of the moon prior to 4.0 aeons. Some consistencies in U-Pb evolution suggest an event at 4.42 aeons which may represent the age of crustal formation and large scale lunar differentiation (Tera and Wasserburg, 1974).

Tera and Wasserburg (1974) discuss two alternative models for the rate of formation of the lunar crust (Fig. 4.3). In the upper diagrams of Figure 4.3 the rate of formation is shown as a function of time while the lower figures show the percentage of the final crust as a function of time. In the first model the crust evolves rapidly such that within 150 m.y. 80 per cent of the final crustal mass is formed. Subsequent to 3.9 aeons only a small addition is made to the crust as the mare basalts are extruded. Two cases are shown in the second model, both involving a uniform rate of crustal growth. In one case the mare basalts are considered a major contributor to the crust, in the other they play a minor part. The second view, in which the basalts are a minor contributor, is more realistic in view of the seismic data presented in Chapter 1.

No conclusive evidence is available to support either model largely because of the lack of quantitative data concerning events in the first 500 m.y. of lunar history. Tera and Wasserburg (1974) prefer, from a schematic viewpoint in the light of the U-Th-Pb systematics, a model in which crust is formed rapidly at first. That is, about 60 per cent of the crust is formed in the first 200 m.y. This fast growth phase would be followed by less intense growth such that about 30 per cent more crust is added in about 3.9 aeons.

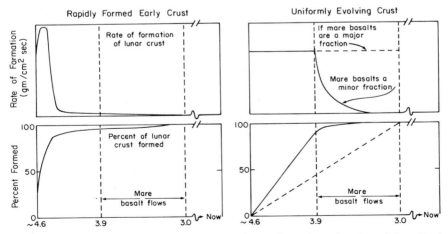

Fig. 4.3. The rate of formation and the percentage of crust formed as a function of time. On the left the crust is assumed to have formed rapidly and early in the moon's history whereas the diagram on the right depicts a situation where the crust evolved at a uniform rate up to 3.9 aeons ago. Mare basalts are assumed to be a minor contribution to the crust (from Tera and Wasserburg, 1974; *Proc. 5th Lunar Sci. Conf., Suppl. 5, Vol. 2,* Pergamon).

The suggestions put forward by Tera and Wasserburg (1974) on the basis of their isotopic data are also consistent with gravity data (Fig. 4.4)

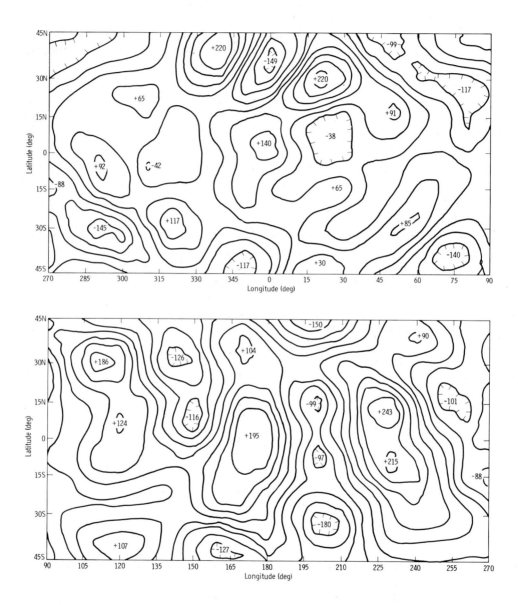

Fig. 4.4. Lunar gravity field; nearside at top, farside at bottom (from Ferrari, 1975; Copyright 1975 by the American Association for the Advancement of Science).

(Muller and Sjögren, 1968; Ferrari, 1975). The lunar highlands and the large mountain ranges surrounding the circular mare are all close to, or at, isostatic equilibrium. For example the Apennine Mountains on the southeastern edge of Mare Imbrium stand 7 km above the mare basalt surface but have a positive anomaly of only +85 mgal. This implies that the crust of the moon was still hot and mobile through the period of mare excavation up to at least 3.85 aeons when Mare Orientale the youngest circular mare, was excavated (Schaeffer and Husain, 1974).

The mare on the other hand show almost no evidence of isostatic compensation such that beneath Mare Imbrium there is an anomaly of +220 mgal. There are similar but somewhat smaller anomalies, or "mascons" as they have been frequently called, beneath most of the mare surfaces (Muller and Sjögren, 1968). The mascons appear to be near surface features and most likely relate to the mare basalt fill (Phillips *et al.*, 1974; Brown *et al.*, 1974). This implies that some time between the excavation of the mare and their flooding by basalt the lunar crust became cold and rigid enough to support the uncompensated mass of mare basalt. The crust must therefore have been essentially in its present state shortly after 3.85 aeons when the last circular mare was excavated and at least sometime between 3.96 and 3.16 aeons when the mare were flooded.

Lunar Stratigraphy

Craters are ubiquitous on the lunar surface (Fig. 4.5). From pits as small as microns on the surface of lunar soil particles they range upward in size to giant craters the size of the Imbrium Basin (Fig. 4.6). The recognition that the large craters are of impact origin and are surrounded by a radially disposed and sculptured ejecta blanket is basic to lunar stratigraphy.

Gilbert (1893) was the first to realize that a circular mare basin, the Imbrium Basin, was impact generated and surrounded by ejecta and that this event might be used to construct a geologic history of the moon. Similarly, Barrell (1927) discussed superposition on the moon as a means of establishing a stratigraphy. Later Spurr (1944-1949) and Khabakov (1960) proposed a series of lunar time divisions. Spurr's divisions were based on structural, igneous and depositional events while Khabakov chose to use topographic differences.

The basis for the present lunar stratigraphy was established in 1962 by Shoemaker and Hackmann although many changes have been made since. Shoemaker and Hackmann based their study on an area to the south of Mare Imbrium in the vicinity of the crater Copernicus where stratigraphic relations

Fig. 4.5. The intensely cratered lunar farside and a portion of the nearside including Mare Crisium (NASA photo AS16-3028)

among the various ejecta blankets are clear (Fig. 4.7). They were able to show that the crater Copernicus was surrounded by an ejecta blanket which consisted of three "facies": a hummocky facies closest to the crater then a radial facies, and finally ray streaks with secondary craters. These three facies were seen to overlie craters such as Eratosthenes and Reinhold which in turn were imposed on the surface of Mare Procellarum. Finally, mare materials were seen to have flooded the Imbrium Basin and to overlie areas of Imbrium ejecta. In all Shoemaker and Hackmann identified five stratigraphic units, four of which they call systems. In order of decreasing age these are (1) pre-Imbrian, (2) Imbrian, (3) Procellarian, (4) Eratosthenian and (5) Copernican. This basic system has been maintained to the present although changes have been made due to a redefinition of some units. Most of the changes are concerned with the definition of rock and time units and their recognition beyond type areas.

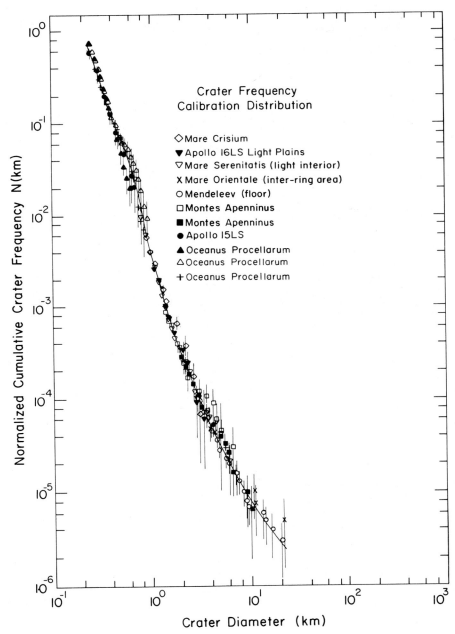

Fig. 4.6. Cumulative size-frequency distributions of all investigated crater populations normalized to the frequency of Mare Serenitatis Light Interior. The solid line represents the polynomial approximation of the calibration size-distribution (after Neukum and König, 1975).

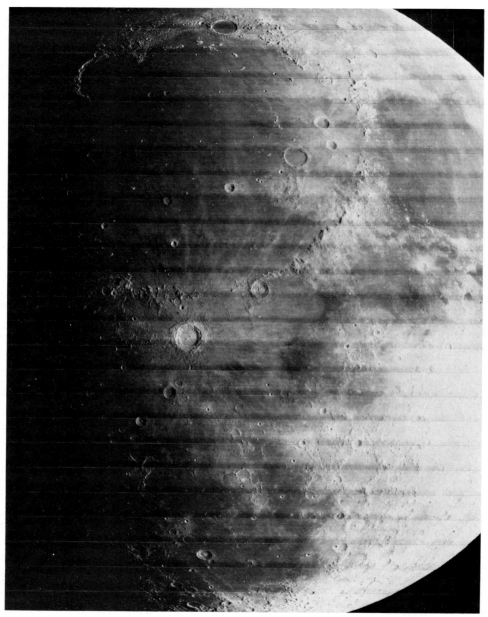

Fig. 4.7. The classic area used by Shoemaker and Hackmann (1962) in establishing the first lunar stratigraphy. The area lies on the southern margin of Mare Imbrium. The arcuate mountain chain at top right forms the margin of the Imbrium Basin. The large rayed crater to the left is Copernicus, the smaller unrayed crater is Eratosthenes. Copernicus is approximately 90 km in diameter (NASA Photo Lunar Orbiter IV-M121).

Stratigraphic Units

A lunar stratigraphy, by necessity, must be based on remotely observable properties of the lunar surface. The units are mapped by outlining areas of similar topographic morphology or albedo or combinations of such properties that can be studied remotely. The stratigraphic significance of such units is a matter of judgment based on an understanding of geologic processes on the moon. In the terrestrial situation the most meaningful stratigraphic units have a three-dimensional form. When dealing with lunar units properties are sought that indicate layers of finite thickness or reflect lithology rather than post-depositional processes (Wilhelms, 1970). Topography is by far the most meaningful property that can be used as an indicator of the three-dimensional properties of the underlying rock body. Simple overall geometric form and lateral uniformity or regular gradation of texture offer best indications of lithologic uniformity, although there is a risk that they reflect a common structural or erosional history instead. Stratigraphic units can be confidently identified if their topographic texture appears similar to primary depositional patterns (Wilhelms, 1970). This points out one of the main weaknesses in lunar geologic mapping, which is that it is necessary to make some assumptions about the genesis of the deposits. It also points out the need for understanding cratering mechanics and the associated depositional mechanisms. Flow lobes, flow lineations and hummocky texture coarsening towards the assumed crater source all help determine the continuity and extent of stratigraphic units. Further confirmation can also be obtained where depositional units bury craters and valleys and thickness determinations are possible (Marshall, 1961; Eggleton, 1963).

It is also possible to use secondary age criteria to supplement superposition and transection relations. Particularly useful is the density of superposed craters and the apparent freshness of units (Wilhelms, 1970). In general crater density increases with age and the rims of older craters become more subdued.

The *lunar material unit* has been established as a parallel to the rock-stratigraphic unit on earth. The lunar material unit has been defined as "a subdivision of the materials in the moon's crust exposed or expressed at the lunar surface and distinguished and delimited on the basis of physical characteristics" (Wilhelms, 1970, p. F-11). Lunar material units have similarities with their terrestrial equivalents but are defined separately because they are observed remotely. That is, features such as the third dimension of the material unit and its lithology may not be known. Further it may be determined to some extent upon inferences based upon our understanding of its genesis.

Local lunar material units may be correlated by their relationship to more widespread units such as ejecta blankets surrounding the large circular mare or by using more subjective criteria such as crater density. Such methods have been employed to establish a stratigraphic column of lunar material units which, like its terrestrial equivalent, has been divided into time-stratigraphic units for use in summarizing geologic history (Shoemaker, 1962; Shoemaker and Hackmann, 1962; Wilhelms, 1970). The major time-stratigraphic units have been called "systems" and their subdivisions "series" as is the terrestrial convention. Similarly the corresponding geologic time units are periods and epochs respectively. The most recent comprehensive assessment of lunar stratigraphic nomenclature is that of Wilhelms (1970). According to his evaluation three systems are presently recognized, which are from oldest to youngest Imbrian, Eratosthenian and Copernican (Table 4.1). Materials older than Imbrian have not yet been assigned a system and are simply called pre-Imbrian. The use of these terms is in some ways unfortunate as much of the stratigraphy outside the Imbrian area is in fact older than Imbrian and eventually will require a more detailed subdivision. It may in fact have been better to avoid the old terrestrial time stratigraphic concepts and treat each major basin as a separate entity with regional correlations. The system names currently used are the same proposed by Shoemaker and Hackmann (1962) except that the Procellarian System has been dropped and the mare materials included in the Imbrian System.

Type areas have been established for each system in the regions where lunar stratigraphy was first studied near the craters Copernicus, Eratosthenes and Archimedes (Fig 4.7). Type areas are used out of necessity on the moon to replace terrestrial type sections. The base of the Imbrian System has been defined as the base of the Fra Mauro Formation, the Imbrium Basin ejecta unit exposed at the surface on much of the Apennine Mountains, Carpathian Mountains and the highlands between the craters Copernicus and Fra Mauro (Wilhelms, 1970). Schaeffer and Husain (1974) place the age of the Fra Mauro Formation at the beginning of the Imbrian at 3.95 ± 0.05 aeons. The top of the Imbrian System is the mare materials. As discussed in a following section, basalt samples returned by the Apollo missions range in age from 3.16 to 3.96 aeons. Presumably the younger limit can be used as a reasonable age for the upper boundary of the Imbrian System. A type mare area has not been designated; however, Wilhelms (1970) has suggested the area between the craters Eratosthenes and Archimedes as such.

Larger rayless craters such as Eratosthenes which are superimposed on the mare surface have been assigned to the Eratosthenian System. The ejecta blankets associated with rayed craters and many dark-halo craters have been assigned to the Copernican System because the rays of the crater Copernicus overlie the rayless craters (Wilhelms, 1970). Lunar time-stratigraphic units express only approximate correlations, because although most rayless craters are older than rayed craters there are exceptions (Wilhelms, 1970). Clearly this is a problem related to the relative albedo of the excavated materials and the surface onto which they are ejected as well as to the age of the craters. The moon's crust is obviously inhomogenous even on a relatively small scale. The absolute age of the Eratosthenian System has not yet been established, but, Eberhardt *et al.* (1973) have suggested that certain unique glass particles found in the lunar soil are Copernican in origin. These glass particles suggest an age of 900 m.y. for the base of the Copernican System.

Major Basin Stratigraphy

A number of large complex circular structures which have been flooded by mare basalts are readily apparent on the lunar nearside. They generally consist of an inner basin and several outer concentric troughs separated by raised, sometimes mountainous, rings (Hartmann and Kuiper, 1962; Baldwin, 1963; Wilhelms, 1970; Howard *et al.*, 1974). These large ringed basins are not confined to the lunar nearside, but are simply more obvious on the nearside due to the basaltic flooding. At least 43 basins larger than 220 km diameter occur on the lunar surface (Fig. 4.8) (Stuart-Alexander and Howard, 1970). These basins appear to be uniformly distributed over the lunar surface (Howard *et al.*, 1974). Many of the farsided basins appear to be extremely old; some of them eroded to the point where they are barely recognisable (El-Baz, 1973).

The circular multiringed basins are by far the largest structures on the moon and tend to dominate both its stratigraphy and structure. The excavation of a ringed basin is a major event and results in the destruction of much of the previous stratigraphic record. Features such as freshness of structures surrounding the basin and crater density on the ejecta blanket have been used to establish the relative times of formation of several of the major nearside basins (Hartmann and Wood, 1971; Hartmann, 1972; Stuart-Alexander and Howard, 1970; Wilhelms, 1970). The relative ages of some of these basins is shown in Table 4.2.

There has recently been considerable debate as to the distribution of these major events in time. The formation of large multiringed basins was clearly terminated prior to the major flooding of the lunar nearside by the mare basalts. The question then is, were these events unique in some way

TABLE 4.1

Composite lunar stratigraphy for seven lunar nearside basins. The columns are arranged in order from east to west across the lunar nearside. Where available, suggested radiometric ages are included beside each stratigraphic unit. The stratigraphy of each basin is initiated by the ejected material from that basin. Not all basin-contemporaneous units have formal stratigraphic names (after Wilhelms, 1970).

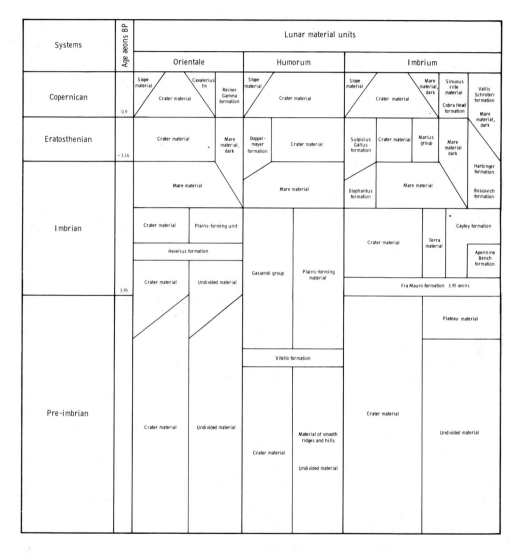

that they occurred in a relatively short period of time, or are the events for which we see a stratigraphic record simply the last of many events which obliterated the earlier record?

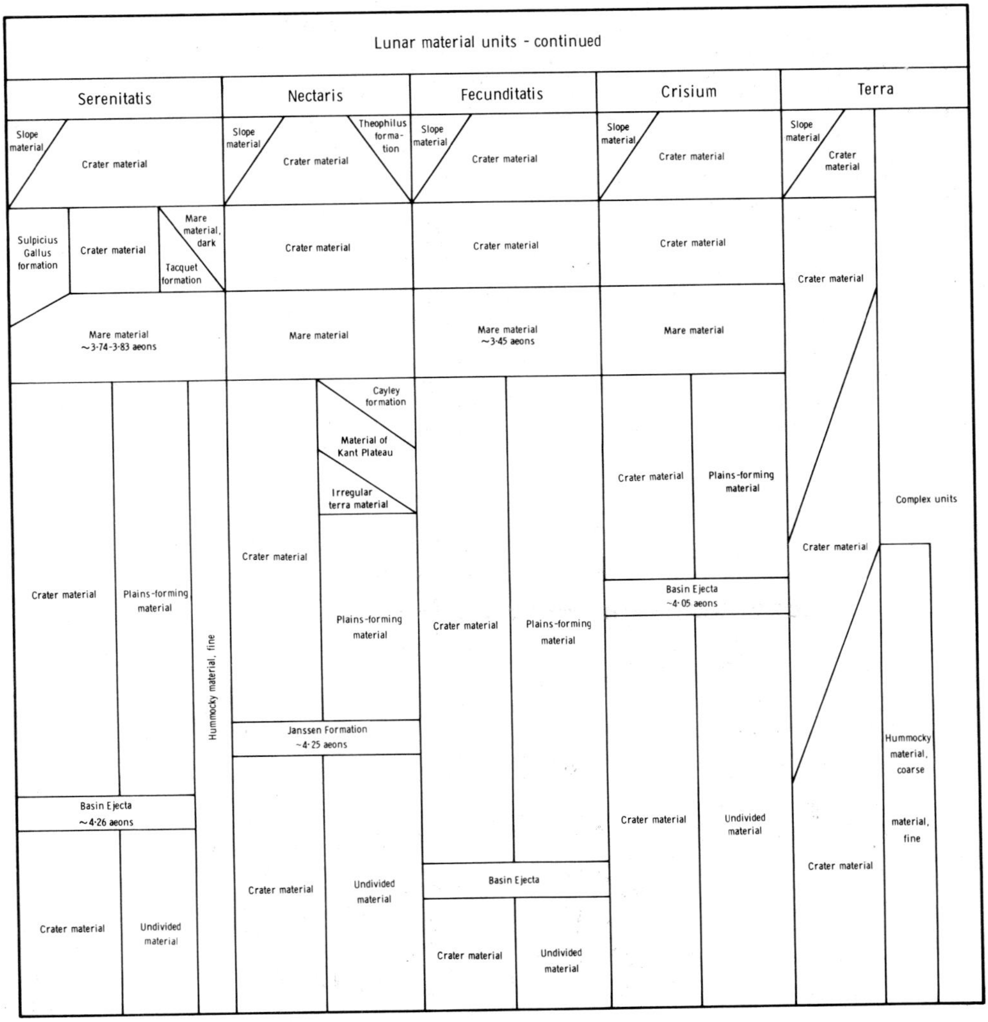
Lunar material units - continued
Serenitatis
Nectaris
Fecunditatis
Crisium
Terra
Slope material
Crater material
Theophilus formation
Sulpicius Gallus formation
Mare material, dark
Tacquet formation
Mare material ~3·74-3·83 aeons
Mare material
Mare material ~3·45 aeons
Cayley formation
Material of Kant Plateau
Irregular terra material
Plains-forming material
Hummocky material, fine
Janssen Formation ~4·25 aeons
Basin Ejecta ~4·26 aeons
Basin Ejecta ~4·05 aeons
Basin Ejecta
Undivided material
Complex units
Hummocky material, coarse
material, fine

Tera *et al.* (1973, 1974a, 1974b, 1974c) have investigated the first hypothesis and have argued that there was a terminal lunar cataclysm at approximately 3.95 aeons ago. The cataclysm is suggested to have extended

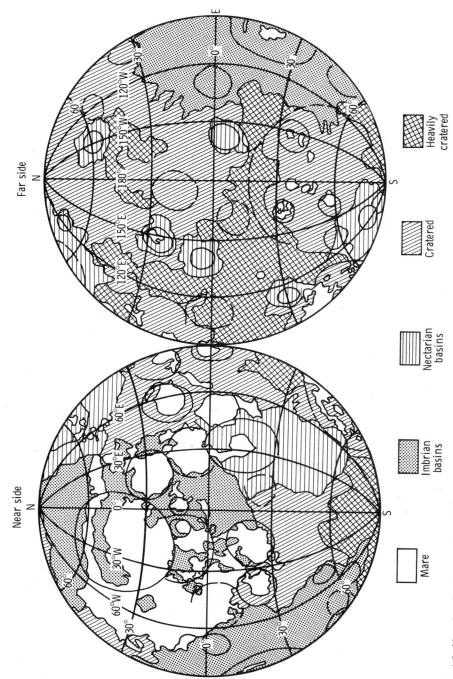

Fig. 4.8. Map showing the location of 43 multiringed basins in relation to the lunar highlands and mare. Imbrian topography is distinguished from older Nectaris basins. Highland areas beyond identified major basin ejecta blankets are subdivided in terms of crater density (from Howard, K.A., Wilhelms, D. and Scott, D.H., Rev. Geophys. Space Phys., 12: 309-327, 1974; copyrighted by American Geophysical Union).

TABLE 4.2

The size and location of some of the more prominent lunar basins (modified after Stuart-Alexander and Howard, 1970). The basins are listed in approximate order of increasing age.

Name	Location		Diameter (km)	Age (aeons)
	Longitude	Latitude		
1. Orientale	−95	−20	900	3.85 ± .05
2. ——	130	−70	300	
3. Imbrium	−19	37	1250	3.95 ± .05
4. Crisium	59	17	450	4.05 ± 4.20
5. ——	−129	3	490	
6. ——	−158	−3	450	
7. Moscoviense	145	25	460	
8. Bailly	−69	−67	310	
9. ——	141	5	330	
10. Humorum	−39	−24	430	
11. Nectaris	34	−16	840	4.25 ± .05
12. ——	160	−53	300	
13. ——	165	−35	370	
14. near Schiller	−45	−34	350	
15. ——	−148	58	300	
16. Grimaldi	−68	−5	430	
17. Serenitatis	19	26	680	4.26 ± .02
18. ——	−153	−35	480	
19. ——	−98	35	320	
20. Humboltanium	81	58	640	
21. Pingre	−79	−56	300	
22. Smythii	84	−3	370	
23. Fecunditatis	51	−3	480	
24. ——	130	−78	370	
25. W. Tranquillitatis	27	9	550	
26. E. Tranquillitatis	38	11	500	
27. Nubium	−17	−19	750	
28. ——	162	−11	480	
29. Australe	90	−45	900	

for approximately 200 m.y. from about 3.8 to 4.0 aeons and resulted in global impact metamorphism. The reason for this suggestion is readily apparent from Figure 4.9 where it can be see that radiometric ages are concentrated at between 3.95 and 4.0 aeons. Husain and Schaeffer (1975)

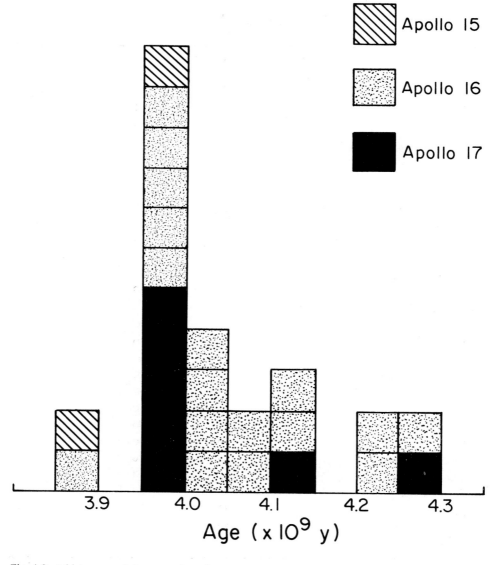

Fig. 4.9. A histogram of the ages of Apollo 15, 16 and 17 breccia samples (from Husain and Schaffer, 1975).

and Schaeffer and Husain (1974) have taken the second view of the origin of the large multiringed basins. They have carefully evaluated the lithology of the breccia samples used in the radiometric analyses and the relative importance of each major basin ejecta blanket at the Apollo landing sites. They have found that ejecta from the Imbrium Basin event, that is, the Fra Mauro Formation for the most part, dominates at all of the Apollo landing sites (Fig. 4.10). Consequently they have concluded that the dominant ages in the 3.95 to 4.0 aeon range simply date the Imbrium event. The Imbrium ejecta blanket is thinnest at the Apollo 16 and 17 sites and it is at these two sites that a number of radiometric ages exceeding 4.0 aeons have been found. As a result of this stratigraphic and radiometric age analysis they have been able to associate absolute ages with several of the major basin forming events (Table 4.2 and Fig. 4.10). They conclude that the era of basin formation extended over many hundreds of millions of years of lunar time up to the excavation of the Orientale Basin. This is consistent with the identification of very old and poorly defined basins on the lunar farside (Stuart-Alexander and Howard, 1970; El-Baz, 1973; Howard *et al.*, 1974). Some of these basins are so badly eroded that they only appear as depressions in the lunar crust and are only recognisable in laser altimeter data. They are undoubtably older than any of the basins listed with radiometric ages in Table 4.2.

Pre-Imbrian Stratigraphy

The pre-Imbrian includes, by definition, all units older than the Imbrium Basin ejecta blanket, that is the Fra Mauro Formation. In terms of geologic time the deposition of the Fra Mauro Formation was essentially instantaneous and it, or at least its lower contact, can be regarded as a time plane. As previously discussed the excavation of the Imbrium Basin occurred approximately 3.95 aeons ago. The use of the Fra Mauro Formation as a time plane and the separation of the Imbrian System from pre-Imbrian units is in many ways an accident both of man and nature. The Imbrium Basin was next to the last of the multiringed basins to be excavated. Consequently, its general features are still clearly discernable in the topography. The Imbrium Basin is also clearly visible from the earth, in contrast to the younger Orientale Basin, only half of which is visible on the lunar nearside. Because almost all of the original stratigraphic studies were made using earth based telescopic photographs the Imbrium area with its clearly defined stratigraphic relations was a prime candidate for study. If the original stratigraphic studies had not been made until after the moon had been photographed from orbit the definition of many time stratigraphic terms may have been

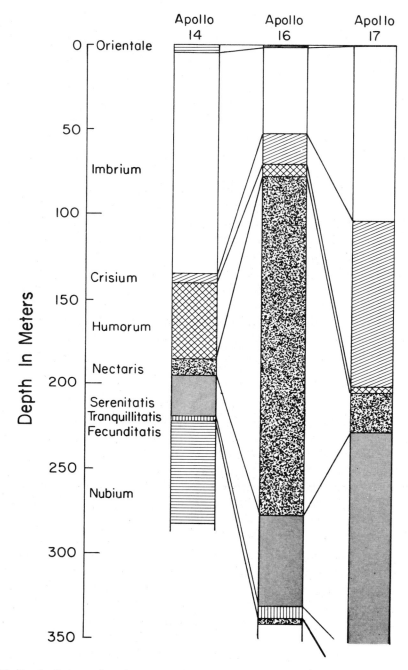

Fig. 4.10. Depth of ejecta, from large basin forming events, at the Apollo 14, 16 and 17 sites (from Schaffer and Husain, 1974).

entirely different.

Most of the pre-Imbrian stratigraphy mapped on the lunar nearside is associated with the older multiringed basins and is of the same age as or younger than the basins but older than the mare material filling the basins. There are thus a large number of stratigraphies, some of which are illustrated in Table 4.1, which are similar to the relationship of the Imbrian units to the Imbrium Basin. Clearly pre-Imbrian stratigraphy will ultimately be divided into many more time-stratigraphic units as the moon's stratigraphy is studied in more detail. The ejecta blanket associated with multiringed basin could be used in the same way as the Fra Mauro Formation to define the time planes subdividing pre-Imbrian time.

Beyond the limits of the Fra Mauro Formation and the Imbrian sculpture the age of material units can not be determined confidently relative to the base of the Imbrian System (Wilhelms, 1970). The top of the Imbrian System is clearly defined in many areas because the mare basalts are wide spread. Frequently material units can be dated in relation to the mare material, either by superposition or by crater density. Consequently some units can not be confidently dated except to say that they are at least Imbrian and possibly pre-Imbrian.

The lower portion of all of the major basin stratigraphies in Table 4.1 are shown as crater material. This broad designation has been established to distinguish very old subdued and degrading craters from the younger Eratosthenian craters. These old crater materials are overlain by the basin-contemporaneous ejecta units which are genetically equivalent and probably lithologically similar to the Fra Mauro Formation. The relative age for some of these units is shown diagramatically in Table 4.1 as are their possible radiometric ages. It should perhaps be pointed out that while the ages given in both Tables 4.1 and 4.2 are absolute their association with a particular multiringed basin is based on, and limited by, stratigraphic interpretations. The only event dated with any real confidence is the Imbrium event due to the fact that Apollo 14 sampled the Fra Mauro Formation directly.

Another series of unnamed pre-Imbrian (and Imbrian) units common to each major basin stratigraphy are the so-called plains forming materials (Wilhelms, 1970). The material units form smooth horizontal surfaces with a high albedo that resemble the Apennine Bench and Cayley Formations of Imbrian age. The last two units are discussed in the following section and in Chapter 5. These units occur over most of the lunar surface generally in the depressions or troughs around each multiringed basin and on the shelf between the inner basin and the first high mountain ring. The light plains units embay the rugged terrain of the basin-contemporaneous units and generally are not cut by the same faults which transect the circum-basin materials. The implication is that the light-plains material units are younger than the ejecta of the associated multiringed basins. The genesis of the plains materials is not well understood and because the stratigraphic relations with the morphologically similar Imbrian units is not known they are generally regarded as either pre-Imbrian or Imbrian in age. Further crater materials occur superimposed on the basin-contemporaneous units. These crater materials are clearly younger than the basin ejecta but older than the mare fillings. The craters range in age across the Imbrian—pre-Imbrian boundary.

The pre-Imbrian, despite its lack of subdivision, is in fact a relatively short time period ($\approx$550 m.y.) compared to the Imbrian and Eratosthenian and is comparable in time available to the post-Cambrian period on the earth. When seen from the point of view of the potential for subdividing the pre-Imbrian time period we quickly realize that ultimately this very early period of the moon's history will be better understood than either Proterozoic or Archaean on the earth. In fact, in some ways, the ancient lunar stratigraphy is likely to be better understood ultimately than the post-Imbrian period simply because most of the post-Imbrian history of the moon is only recorded in the very thin lunar soil where the time resolution may not be good.

The Imbrian System

The Imbrian System has not been formally defined. However, Wilhelms (1970) who produced the most complete summary of lunar stratigraphy, has stated that the base of the Imbrian System is the Fra Mauro Formation and the top is mare material. It thus includes all of the time between the excavation of the Imbrium Basin and the ultimate filling of the basin by basalt flows. The term "Imbrian System" was originally applied to the hummocky blanket surrounding Mare Imbrium (the present Fra Mauro Formation) by Shoemaker and Hackmann (1962). Since this time the Imbrium System has gone through a large number of modifications mostly on a relatively informal basis (Wilhelms, 1970; Shoemaker et al., 1962; Hackmann, 1962; Marshall, 1963). Material units included within the Imbrian System are listed in stratigraphic order in Table 4.1.

The Fra Mauro Formation

The Fra Mauro Formation forms the basal unit of the Imbrian System (Fig. 4.11.). The term was first used by Eggleton (1964, 1965) although the formation was not formally defined until much later (Wilhelms, 1970). The type area lies between latitudes 0° and 2°S and longitudes 16° and 17°30'W to

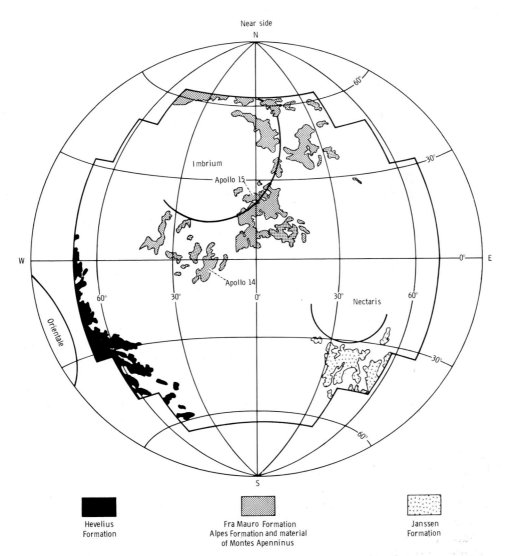

Fig. 4.11. Nearside of the moon showing textured basin ejecta (from Howard, K. A., Wilhelms, D. E. and Scott, D. H., *Rev. Geophys. Space Phys.*, *12*: 309-327, copyrighted American Geophysical Union).

the north of the pre-Imbrian crater Fra Mauro after which it is named. Eggleton isolated two members (facies) on the basis of topography, one with a hummocky surface, the other smooth. The type area is in the hummocky facies where the surface consists of abundant closely-spaced hummocks (Fig. 4.12). The hummocks are low rounded hills 2-4 km across and are approximately subequidimensional although there is a tendency towards a

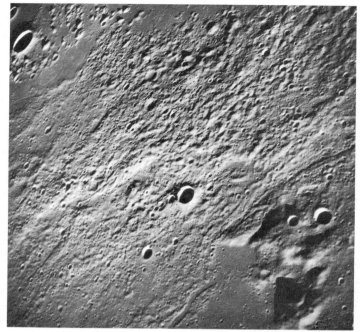

Fig. 4.12. The Fra Mauro Formation close to its type area south of Mare Imbrium. Note the hummocky nature of the formation in the north and the gradual increase in smoothness further south (NASA Photo AS12-52-7597).

north-south elongation. The regular nature of the topography suggests that it is a single depositional unit which probably consists of ejecta resulting from the excavation of the Imbrium Basin. Samples of the Fra Mauro Formation returned by the Apollo 14 mission support this view. They consist of poorly-sorted breccias which are metamorphosed to varying degrees. Chapter 5 is devoted largely to a discussion of these lithologies.

In the vicinity of the type area the formation averages 550 m in thickness although lateral variations in thickness are considerable due to the variable relief of the pre-Imbrian terrain (Eggleton, 1963). Other local estimates of the thickness of the Fra Mauro Formation have been made by McCauley (1964) and Eggleton and Offield (1970) who, like Eggleton (1963) used the degree of burial of pre-Imbrian craters in relation to crater size to make the estimates. Kovach *et al.* (1971) used active seismic data to determine the thickness of the Fra Mauro Formation at the Apollo 14 site and concluded that it is between 46.5 and 84.5 m thick. McGetchin *et al.* (1973) made estimates of the thickness of the Fra Mauro Formation using empirically derived formulae based on cratering models. Their estimates of the radial thickness variations are shown in Figure 4.13. In general this empirical model agrees well with all other available estimates and is

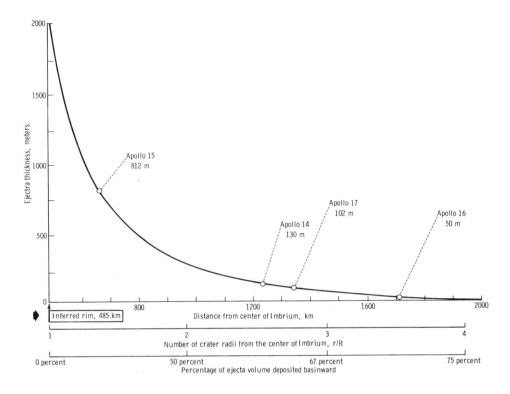

Fig. 4.13. Radial thickness variation of the Fra Mauro Formation (from McGetchin *et al.*, 1974).

consistent with a base surge depositional model (see Chapter 3). From Figure 4.13 it can be seen that the formation thins dramatically away from the basin and that 50 per cent of the ejected material lies within one crater (basin) radius of the inferred rim of the Imbrium Basin.

Exposures of the Fra Mauro Formation are relatively limited in extent beyond the type area by mare basalt flooding and light-plains units which may be Cayley Formation (see following section). Correlations beyond the type area are difficult but the formation is believed to extend all around the Imbrium Basin (Wilhelms, 1970) which is again consistent with cratering models.

Beyond the type area of the Fra Mauro Formation away from Mare Imbrium the hummocks on the surface of the formation gradually disappear and it grades into the "smooth" member (Fig. 4.12). The surface of this member has very low ripple-like hummocks or sinuous ridges. In general this member occurs farther from the basin than the hummocky member. However, local occurrences of the smooth member are found on topographic highs closer to the basin. It is difficult to visualize the transportational mechanism involved in an event of the magnitude of the Imbrium event. However, the regular nature of the hummocky terrain and the fact that the hummocky terrain dies out farther from the basin suggests transport by a giant base surge, at first in supercritical flow and then becoming subcritical as the flow is decelerated by loss of momentum due to gas loss. This is also consistent with the occurrence of the smooth member on topographic highs which presumably caused a momentary local braking of the flow. If base surge was the primary mechanism of transport the regular hummocks may be related to antidune structures. The 2-4 km dimensions of the hummocks in the type area would imply flow velocities of the order to 23 to 32 m s^{-1} which is consistent with base surges of a much smaller scale formed during terrestrial volcanic eruptions and experimental explosions. Mean velocities of this order of magnitude would require about 20 hr to complete the deposition process. The textural evidence from the breccias that suggests base surge is discussed in more detail in the following chapter.

The Cayley and Apennine Bench Formations

Plains-forming materials with a high albedo occur abundantly on the lunar nearside. Stratigraphic relationships in the northern and central part of the lunar nearside suggest that these plains are younger than the Fra Mauro Formation and the associated Imbrian sculpture, but older than the local mare basalts (Wilhelms, 1970). The Apennine Bench Formation consists of light plains-forming materials near the crater Archimedes and was defined originally by Hackmann (1964, 1966). Similar high albedo plains occur in a circum-Imbrium trough in the vicinity of the crater Cayley (Wilhelms, 1965; Morris and Wilhelms, 1967). Morris and Wilhelms defined this material unit as the Cayley Formation (Fig. 4.14). The Cayley Formation appears independent of Imbrium sculpture, and is embayed by mare basalts from Mare Tranquillitatis. It occurs in topographic lows and generally has a smooth and horizontal surface. The contacts of the Cayley Formation with surrounding units are extremely variable ranging from sharp to gradational. Locally many occurrences of Cayley Formation merge into areas of subdued topography without a detectable change in albedo. Milton (1968) and Wilhelms (1968) refer to these material units as the "hilly" member of the Cayley Formation.

Prior to the Apollo landings the Cayley Formation was generally believed to be volcanic in origin, and to consist either of flows or pyroclastic materials (Wilhelms, 1970). The Apollo 16 landing site (Fig. 4.14) was selected in part to determine the nature of the Cayley Formation; it would have been particularly important if it had in fact been the result of late stage volcanism. As a result of this mission it can be concluded that the Cayley

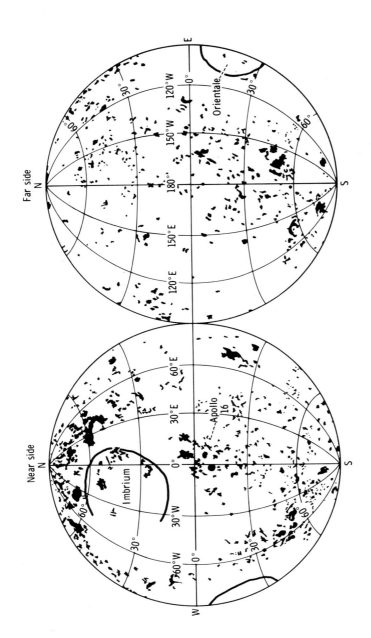

Fig. 4.14. Distribution of Imbrian age plains (Cayley Formation). The Apollo 16 landing site is indicated. Plains tend to be concentrated outside the ejecta blankets of the Orientale and Imbrium basins (from Howard, K. A., Wilhelms, D. E. and Scott, D. H., *Rev. Geophys. Space Res.*, *12*: 309-327, copyrighted American Geophysical Union).

Formation is not volcanic in origin but consists of impact generated clastic rocks (Warner *et al.*, 1973; Wilshire *et al.*, 1973; Walker *et al.*, 1973; Bence *et al.*, 1973). The problem is which event or events to associate the formation with. There are essentially three theories of its origin (1) it is ejecta from a large basin-forming event possibly Orientale, (2) it is secondary ejecta from a large scale event such as Orientale, or (3) it is locally derived material moving down slope, filling depressions by slumping or possibly as secondary ejecta. Several authors (Hodge *et al.*, 1973; Eggleton and Schaber, 1972; Chao *et al.*, 1973) have argued that the Cayley Formation is extremely widespread (Fig. 4.14) and occurs in areas well beyond the potential source event at Orientale. If McGetchin *et al.*'s (1973) empirical relationship is correct the Orientale ejecta deposits would be extremely thin at many Cayley Formation occurrences, particularly at the Apollo 16 site, and the volumetric requirements for such an extensive unit would be unrealistic in terms of this one event. This last problem is more easily resolved if we appeal to the mechanism proposed by Oberbeck *et al.* (1974) which suggests that the Cayley Formation consists of secondary ejecta produced by Orientale ejecta. This cascade effect could potentially double the volume of available materials. Possibly the strongest evidence in favor of a local origin for the Cayley Formation comes from orbital geochemical data (Hörz *et al.*, 1974). There is no systematic relationship between either Al/Si values determined by the X-ray fluorescence experiment or the Th values determined by gamma ray counting and the mapped distribution of the Cayley Formation (Fig. 4.14). In general the Cayley Formation is similar in composition to the surrounding surficial deposits.

The Orientale Basin

The Orientale Basin (Fig. 4.15) is the only large multiringed basin younger than the Imbrium Basin. The material units surrounding this basin are very similar in their texture and distribution to the Fra Mauro Formation (McCauley, 1964). The inner hummocky facies which extends for about 900 km has not been formally named but is undoubtably genetically equivalent to the hummocky facies of the Fra Mauro Formation. The name Hevelius Formation has been applied to the smooth outer facies.

McCauley (1969) described some very distinctive topography on the hummocky material units surrounding the Orientale Basin. Dune-like structures and braided surfaces appear intermixed on the surface of the unit (Fig. 4.16). The dunes are often more pronounced in front of major topographic obstacles whereas the braids are best developed on level upland surfaces. These features are again very suggestive of bedforms produced during super-critical flow and probably relate to a very large base surge

Fig. 4.15. An orbital view of the Orientale Basin surrounded by its braided and hummocky ejecta blanket. The annular mountain ranges are clearly visible. Mare Orientale is approximately 900 km in diameter (NASA Photo Lunar Orbiter IV-M 187).

flowing outward from the Orientale Basin immediately following its excavation. The topographic relationships suggest that the braided bedforms may develop on flat surfaces at higher velocities, while the dune-like structures occur where local topography has reduced the velocity of the

Fig. 4.16. Braided ejecta materials surrounding Mare Orientale. This area lies about 1100 km southeast of Mare Orientale. The crater Inghirami is about 90 km across and is only partly covered by the swirling braided ejecta (NASA Photo Lunar Orbiter IV-H172).

flow.

The Orientale Basin itself has several concentrically arranged scarps which form two mountain ranges (Fig. 4.15). The inner range is called the Rook Mountains and the outer-most range the Cordillera Mountains. The origin of these regularly arranged concentric scarps has been much debated (Hartmann and Yale, 1968; Dence, 1974; van Dorn, 1968; Baldwin, 1972). The scarps may be related either to shock waves produced during cratering or to tectonic subsidence following cratering. The center of the basin inside the Rook Mountains is flooded by mare basalts. The material unit between the Rook and Cordillera Mountains has been informally called the Montes Rook Formation. It is distinguished by a relatively irregular topography of small, smooth and closely spaced hills. On the insides of the Montes Rook Formation a second material unit with a higher albedo can be distinguished.

These units may be either fall back or slump fractured rim deposits (McCauley, 1968). Due either to great age or mare basalt flooding similar deposits are not readily apparent in other multiringed basins.

The Orientale material units are clearly older than the mare basalts. The lower density of craters on the Hevelius Formation and its hummocky correlative strongly suggest that they are younger than the Fra Mauro Formation. This is supported further by the sharpness and freshness of the basin scarps. If Schaeffer and Husain (1974) are correct in their analysis the Orientale Basin was formed 3.85 aeons ago (Tables 4.1 and 4.2) which places it early in Imbrian time.

The Mare

The dark smooth surfaces with low albedo which form the lunar nearside mare (Fig. 4.17). are obvious even to the casual observer standing on the earth's surface. Perhaps the first important contribution by Apollo to our knowledge of the moon was to verify early suggestions that the mare consisted of basalt flows (Baldwin, 1949; Kuiper, 1954; Fielder, 1963).

Because, when studying the stratigraphy of the moon, we are dealing mainly with material units, Wilhelms (1970) defines mare material as being "dark flat and smooth" and locally having "characteristic ridges and domes." The only other units which could be confused with mare materials are the smooth flat materials similar to the Cayley Formation. The main differences between these units and the mare are related to the interdependent variables of crater density and albedo. These variables simply reflect age — the mare material being much younger.

Baldwin (1963) and Mutch (1972) both recognized the presence of old lava flow fronts on the mare surface (Fig. 4.18), particularly in southwestern Mare Imbrium. It thus became clear that the mare were similar to terrestrial flood basalts such as the Deccan Traps and that the mare were probably underlain by an extensive stratigraphy. When Apollo 15 landed beside Hadley Rille in southern Mare Imbrium this view was further supported by observations of massive layering in the opposite rille wall (Fig. 4.19).

The sequence of well-defined flow fronts visible under low-angle lighting conditions in southwestern Mare Imbrium has received detailed attention (Schaber, 1973), and has provided considerable insight into the mare filling process. These late stage basalt flows are believed to be Eratosthenian in age. The age is based on crater counts and in light of past problems should be treated with caution until the validity of the method is more firmly established.

The eruption of the flows occurred in at least three phases (Fig. 4.20). The source of the youngest lava appears to be a 20 km long fissure close to

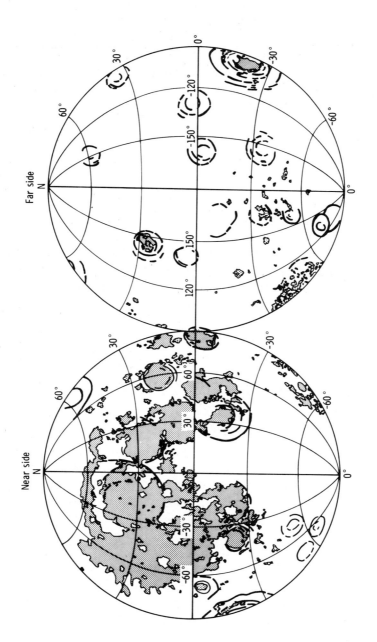

Fig. 4.17. Distribution of mare (shaded) and large circular basins. The highest mountain ring of each is shown by a heavy line; secondary mountain rings are shown by the light lines (after Stuart-Alexander and Howard, 1970).

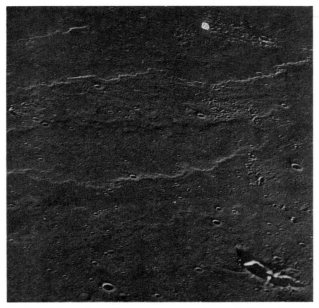

Fig. 4.18. A well-defined lava flow front on the surface of Mare Imbrium (NASA Photo AS15-96-13023).

Fig. 4.19. Massive horizontally bedded units in the wall of Hadley Rille (NASA Photo AS15-89-12100).

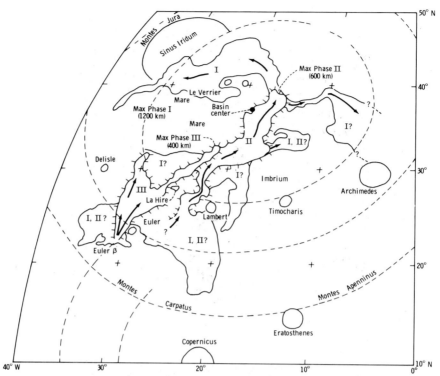

Fig. 4.20. Distribution of late-stage mare basalt flows in the Imbrium Basin (from Schaber, 1974; *Proc. 4th Lunar Sci. Conf., Suppl. 4, Vol. 1*, Pergamon).

the crater Euler at 22°50' N and 31°20' W near the southwestern edge of Mare Imbrium. Possibly significant is the fact that this source lies close to a seismically active region (Latham *et al.*, 1974) and in an area where the Imbrium Basin appears to intersect an older poorly-defined basin. This suggests that the migration of lava to the surface may be controlled by faults associated with the major multiringed basins. The lavas from the three eruptive phases extended 1200, 600, and 400 km respectively indicating a gradual decrease in magma volume with time. These flows accounted for a volume of at least 4×10^4 km^3 over an area of 2×10^5 km^2. The figures are of a similar order of magnitude as terrestrial flood basalts.

Individual lava flows range from 10 to 63 m in thickness and flowed down slopes with a gradient between 1:100 and 1:1000. The flows surmounted local uphill slopes of at least 0.5° but locally were ponded behind mare ridge crests. Leveed channels developed at the ridge as the lava overflowed some ridges. Deep leveed channels are also present along the center of some flows and braided channels have been seen near source vents (Schaber, 1973). Lava channels are common on the earth but the lunar lava channels are several orders of magnitude larger. The extreme distances to which the flows travelled on the lunar surface probably related to a rapid rate of extrusion of the lava with low melt viscosity playing a secondary role. It could thus be construed that the mare are underlain by a well-stratified complex sequence of basaltic flows.

Mare Surface Features. The extreme smoothness of most of the mare surfaces make the few topographic features all the more pronounced. While many minor topographic features have been described there are two large scale features apparent in most mare; rilles and wrinkle ridges.

The use of the term "rille," which was originally helpful in terms of classifying lunar topographic features, is perhaps, in the long term, unfortunate. There are three general types of rille which relate to at least two different mechanisms; straight, arcuate and sinuous (Schubert *et al.,* 1970; Schumm, 1970; Howard *et al.,* 1972; Oberbeck *et al.,* 1972). Straight and arcuate rilles appear to be genetically related (Figs. 4.21 and 4.22). They may be many kilometers long and several kilometers across. Typically these features transect both mare and highland surfaces and there appears little

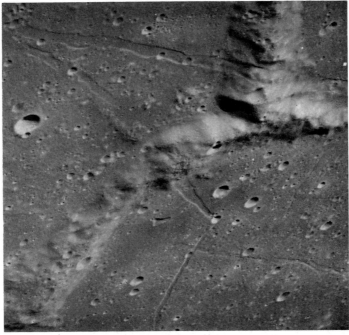

Fig. 4.21. Graben-like straight rilles transecting mare and highland surfaces (NASA Photo AS14-73-10115).

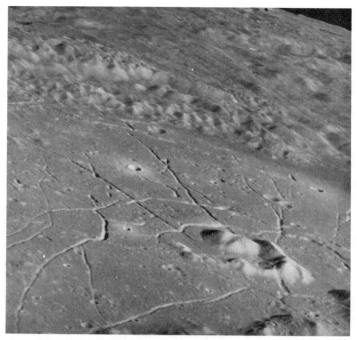

Fig. 4.22. Straight and arcuate rilles in combination (NASA Photo AS15-93-12642).

reason to suggest that they are anything but fault controlled graben-like structures (McGill, 1971; Mutch, 1970; Young *et al.,* 1973).

Sinuous rilles, as the name suggests, look like meandering river channels (Fig. 4.23) which, in the past, led some authors to suggest that they were eroded by water (Lingerfelter *et al.,* 1968; Schubert *et al.,* 1970). The majority of sinuous rilles occur along the edge of the mare, sometimes running along the highland front but never crossing into the highlands (Fig. 4.24). In cross section the rilles may be V-shaped but generally their profiles are more U-shaped and give the impression of having been smoothed or subdued by subsequent erosional processes (Fig. 4.25 and 4.26). In some cases rilles have flat floors and inner sinuous channels (Fig. 4.27). Longitudinal profiles of sinuous rilles generally slope in the same direction as the regional slope (Gornitz, 1973) (Fig. 4.28). However, there are some unexplained anomalies which are inconsistent with open-channel fluid flow. For example, the source of Prinz I Rille is deeper than its termination, perhaps suggesting that the crater was drained by an alternate channel. Marius Hills Rille displays a rise where it crosses a wrinkle ridge. The gradient

Fig. 4.23. Sinuous rilles on the Aristarchus Plateau. Most are flat bottomed and some have inner sinuous channels. The large partially flooded crater (Prinz) at top right is ≈40 km in diameter (NASA Photo AS15-93-12608).

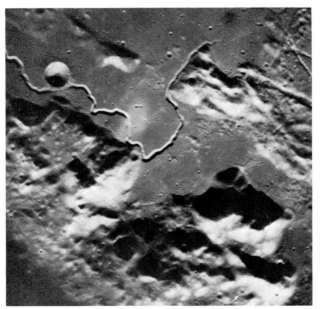

Fig. 4.24. An orbital view of Hadley Rille the sinuous rille visited by the Apollo 15 mission. Note that the rille follows the mare-highland boundary for short intervals but never crosses into the highlands; thus implying a genetic association with the mare filling (NASA Photo AS15-94-12813).

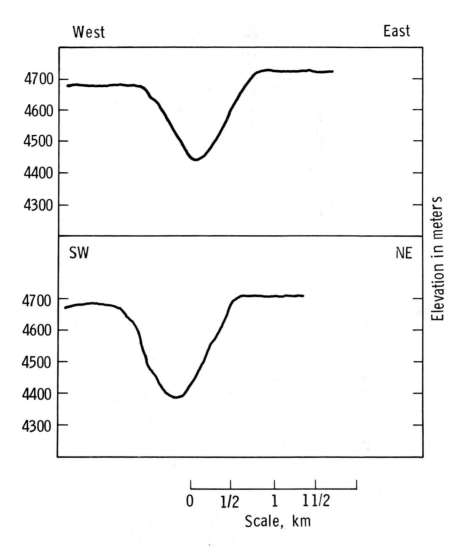

West   East

East Elevation in meters

SW   NE

0   1/2   1   1 1/2

Scale, km

Fig. 4.25. Cross sections of Hadley Rille (from Gornitz, 1973).

of the rilles typically ranges from 0.0224 to 0.0026. These gradients are steeper than some terrestrial rivers crossing their flood plains. The dimensions of some sinuous rilles cover a considerable range but show some internal consistencies. The relationship between depth and width are roughly linear (Fig. 4.29). They have walls which slope between 10° and 23° with an average slope of 17° (Gornitz, 1973). The meander length of the rilles is approximately twice the channel width. The radius of curvature of the rille

Fig. 4.26. Hadley Rille viewed from the level of the mare surface. Notice the subdued appearance of the rille and the large numbers of boulders which have rolled to the rille floor (NASA Photo AS15-85-11451).

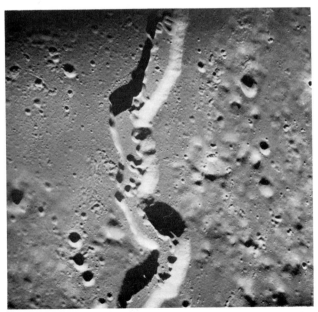

Fig. 4.27. Schroeters Valley, a broad flat-bottomed sinuous rille with an inner channel (NASA Photo AS15-97-13258).

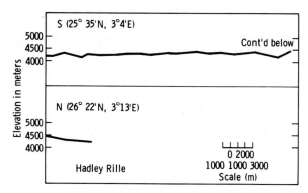

Fig. 4.28. Longitudinal bottom profile of Hadley Rille (from Gornitz, 1973).

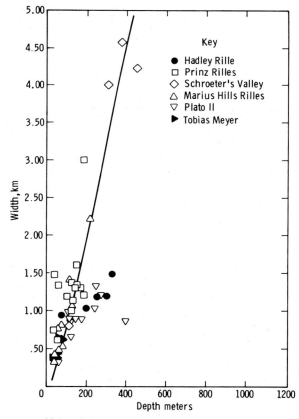

Fig. 4.29. Relation between width and depth of sinuous rilles (after Gornitz, 1973).

meanders is generally less than 1 (Fig. 4.30) in contrast to rivers where it
may be as much as 4.3. No correlation exists between channel width,
meander length or radius of curvature. This is in contrast to terrestrial rivers
but much more in keeping with lava tubes and channels. Sinuous rilles are
thus wide relative to meander length and the radius of curvature is small. The
consensus at the present time is that most sinuous rilles are lava tubes, the
roofs of which have collapsed presumably due to meteoroid bombardment
(Howard *et al.*, 1972). Arcuate depressions are seen in places on the mare
surface following sequentially in a sinuous path (Fig. 4.31). The depressions
may in fact be the partially collapsed roof of a sinuous lava channel or due
to later stage fissure eruptions. Young *et al.* (1973) have pointed out other
evidence of partial bridging of sinuous rilles. The abrupt termination of
many rilles in a downstream direction is also consistent with a lava tube
origin as lava tubes may disappear upon entering lava lakes or pools. Finally
the overall geometry is most similar to terrestrial lava tubes.

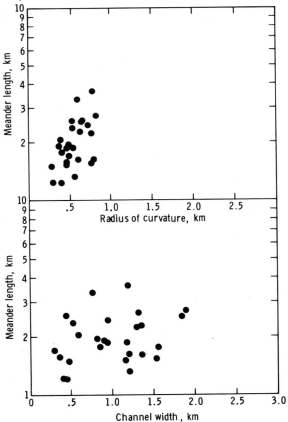

Fig. 4.30. Geometry of sinuous rilles (from Gornitz, 1973).

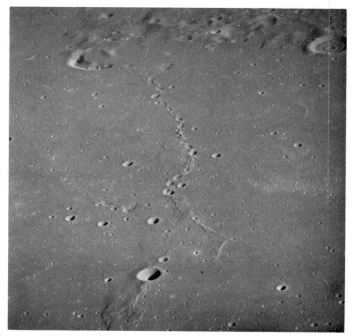

Fig. 4.31. This sinuous line of depressions may indicate the presence of a partially collapsed lava tube or a late stage fissure (NASA Photo AS15-93-12725).

Some rilles have conspicuous levees along their margins. They are generally less sinuous than other rilles and have been interpreted as being open channels on the surface of lava flows (Young *et al.*, 1973). The rilles are large by comparison with terrestrial equivalents and it has been difficult to accept a simple lava tube or channel hypothesis. However, Carr (1974) has carried out a series of simulation studies based on our present understanding of the conditions of the lava during extrusion, and he has shown that such channels may develop very rapidly (Fig. 4.32). The erosion rate is very sensitive to the difference between the lava temperature and the yield temperature. The results of Carr's work imply that large amounts of erosion are possible under conditions of sustained flow and that lava channels may deepen at rates in excess of 1 m per month on the lunar mare.

Sinuous rilles of volcanic origin are presumably of the same age as the surface flows in which they occur. The length of sinuous rilles thus give indications of the dimension of lava flows flooding the mare. Some rilles extend for between 300 and 400 km showing the extreme mobility of the lunar lavas.

Wrinkle ridges (Fig. 4.33) are intimately associated with mare materials. The ridges are generally concentric to the basin but radial ridges also occur.

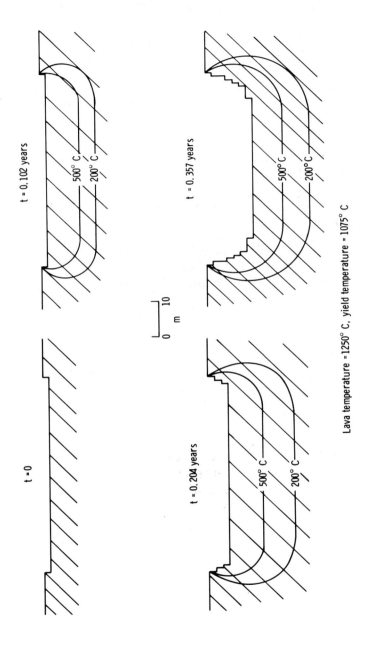

Fig. 4.32. Computer modelling of the growth of a lunar lava channel. Growth is rapid because of the high lava temperatures (after Carr, 1974).

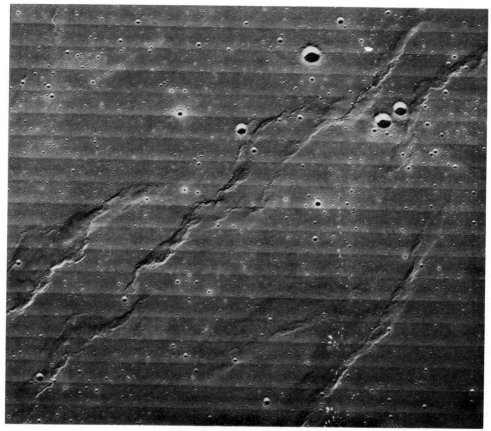

Fig. 4.33. Wrinkle ridges in Oceanus Procellarum (3°50' S, 36° W). The ridges in view are maximum 3 km wide (NASA Photo Lunar Orbiter V-M172).

The ridges may develop as low swells concurrently with the inpouring of the most recent lava flows (Bryan, 1973). At some localities, particularly in Mare Imbrium, lava flows can be seen breaching the ridges and ponding behind ridges (Fig.4.34). The ridges appear to be deformational structures caused by compression (Bryan, 1973). The surfaces of the mare apparently were subsiding at the same time as they were being flooded. The subsidence produced fault scarps and monoclinal folds which resulted in the ponding of the lavas. The lava flows were then followed by more compressions. The close genetic relationship between wrinkle ridges and mare fillings is shown by the fact that the ridges tend to occur at approximately half the distance from the center to the margin of the basin (Fig. 4.35).

   Age and Duration of Mare Volcanism. Thermal activity within the moon appears to have ended, at least in large part, at the time the mare

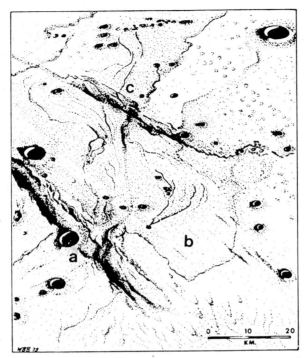

Fig. 4.34. Lava flows and wrinkle ridges on the south side of Mare Imbrium. A flow coming from the lower left crossed a ridge (a) and was ponded (b). The lava broke through the second ridge at (c) (from Bryan, 1973; *Proc. 4th Lunar Sci. Conf., Suppl. 4, Vol. 1,* Pergamon).

basalts were extruded. Tectonism continued into the post-mare filling period as evidenced by the straight and arcuate rilles but there is no evidence of either further volcanism or of any activity that could be interpreted as orogenesis or isostatic compensation. At this point in time the lunar crust appears to have become rigid and the moon became a "dead" planet.

Even though the mare filling was the most recent major activity on the moon, by terrestrial standards mare volcanism is extremely ancient. Radiometric ages provided by two main methods (Rb-Sr and ^{40}Ar-^{39}Ar) indicate a range of between 3.16 to 3.96 aeons for all available Apollo samples (Papanastassiou and Wasserburg, 1971, 1972; Wasserburg *et al.,* 1974). The time period is thus 800 m.y. which by terrestrial standards is longer than all of the time available from Cambrian to present. However, if we view the data on a basin by basin basis (Table 4.3) it is apparent that the continuity of mare volcanism is something of an illusion created by the morphologic continuity of the mare on the lunar surface. Keeping in mind the limited sampling, the mare flooding appears to have occurred for periods of 100 to 400 m.y. at different times and at different locations on the lunar

Fig. 4.35. Wrinkle ridges in Mare Humorum. Mare Humorum is about 350 km in diameter (NASA Photo Lunar Orbiter IV-M137).

surface. From the limited data it is apparent that there are large periods of time between the excavation of the multiringed basins and the flooding of the basins by mare basalt. For example, the Imbrium Basin was excavated 3.95 aeons ago but the oldest Imbrium basin sample is 3.44 aeons old; a time gap of 510 m.y. A similar time gap is apparent fof the Serenitatis Basin. This tends to suggest that the mare basalts were not the direct result of impact melting during basin excavation. Arakani-Hamed (1974) has proposed that the basin forming events did not trigger mare volcanism directly but the

TABLE 4.3

Age of mare basalts (data from many sources).

Mission	Mare	Age Range $(x\ 10^9\ yr)$	Time Period m.y.	Excavation of Mare Basin $(x\ 10^9\ yr)$
Apollo 14		3.95–3.96	100	
Apollo 17	Serenitatis	3.74–3.83	90	4.26
Apollo 11	Tranquillitatis	3.51–3.91	400	4.26
Luna 16	Fecunditatis	3.45		4.26
Apollo 15	Imbrium	3.26–3.44	180	3.95
Apollo 12	Procellarum	3.16–3.36	200	?

thermal balance of the moon was upset resulting in late magmatization and volcanism. This model poses further problems. In the first place it does not explain the absence of lava fillings in the lunar farside basins and second the time interval between basin excavation and filling does not appear to be regular. For example, Tranquillitatis and Fecunditatis are two of the oldest basins — both are older than Serenitatis — and yet they were flooded by basalt at about the same time or later. Overall, however, data are very limited. The density of rilles around the margins of the mare suggests that the most recent volcanic activity may have occurred around mare margins. By design the Apollo missions sampled the mare margins so that highland and mare could be sampled by a single mission. It is consequently difficult to make a definite statement concerning the relationship between basin excavation and flooding except that flooding appears to have occurred some time after excavation. This is also consistent with the pre-flooding cratering history of the mare basins (Baldwin, 1949).

The oldest dated mare basalts are clasts taken from Fra Mauro Formation breccias (Papanastassiou and Wasserburg, 1971). The greatest age found (3.96 aeons) is only slightly older than the Fra Mauro Formation (3.95 aeons) and suggests that mare volcanism began in late pre-Imbrian time. These basaltic fragments may have come from an older basin commonly referred to as "South Imbrium" which is now largely obliterated by Imbrian ejecta (Taylor, 1975). The youngest directly dated mare materials are those returned by Apollo 12 from Oceanus Procellarum. There are indications, however, that the youngest mare materials were not sampled by the Apollo missions. Structures of probable volcanic origin are super-imposed on the mare surface in the Marius Hills area and may be younger than the Procellarum lavas (3.16 aeons). On the basis of crater counts, Schaber (1973) has suggested younger ages for the most recent lavas in southwestern Mare Imbrium. It thus seems that the upper and lower boundaries of the mare materials are considerably more complex than the generalized stratigraphy shown in Table 4.1 would suggest.

Composition of Mare Basalts. The lunar mare were flooded with high-iron basalts with a wide range in composition. Compositional variations are large both within and between sampling sites. In comparison with the earth the lunar basalts at each sampling site are texturally and compositionally much more varied. Mineralogically the most striking feature of the lunar basalts is their extreme freshness. Unlike the earth where water is abundant, there are virtually no secondary minerals due to weathering on the moon unless one regards impact produced glasses as the equivalent to terrestrial weathering products. As might be expected clinopyroxene and plagioclase dominate the basaltic mineralogy with olivine and ilmenite being important in some samples.

Pyroxene is the most abundant mineral in the lunar basalts and frequently forms more than 50 per cent of the rock (Table 4.4). Compositionally the pyroxenes are extremely complex. This complexity is determined largely by the chemistry of the host rock, paragenetic sequence, emplacement history and oxygen fugacity (Bence and Papike, 1972). Single pyroxene crystals frequently show extreme zoning (Fig. 4.36 a,b) and this combined with textural information suggests a definite sequence of pyroxene crystallization: magnesian pigeonite followed by magnesian augite and finally more iron-rich compositions (Dowty *et al.*, 1973). Possibly the best indication of the complexity and subtlety of compositional variations in the pyroxenes can be gained from the detailed study of the Apollo 15 basalt made by Dowty *et al.* (1974) (Fig. 4.36). Pyroxenes from feldspathic peridotite (Fig. 4.36c) contain more magnesium in the extreme range than other lithologies and few of the pyroxenes are iron rich. Olivine gabbros contain less magnesium again in the extreme range followed by olivine-phyric basalts which are lower still. Concurrent with the decreasing magnesium content in the extreme compositions is an iron enrichment.

The plagioclases have a relatively restricted compositional range with extremes of An_{60} and An_{98}. The crystals are frequently zoned although it

TABLE 4.4

Modal mineralogy of mare basalts (values in percentages) (after Taylor, 1975).

	Olivine basalt Apollo 12	Olivine basalt Apollo 15	Quartz basalt Apollo 15	Quartz basalt Apollo 12	High-K basalt Apollo 11	Low-K basalt Apollo 11	High-Ti basalt Apollo 17	Aluminous mare basalt Apollo 14	Luna 16
Olivine	10-20	6-10	–	–	0-5	0-5	5	–	–
Clinopyroxene	35-60	59-63	64-68	45-50	45-55	40-50	45-55	50	50
Plagioclase	10-25	21-27	24-32	30-35	20-40	30-40	25-30	40	40
Opaques	5-15	4-7	2-4	5-10	10-15	10-15	15-25	3	7
Silica*	0-2	1-2	2-6	3-7	1-5	1-5	–	2	–

*Tridymite, cristobalite

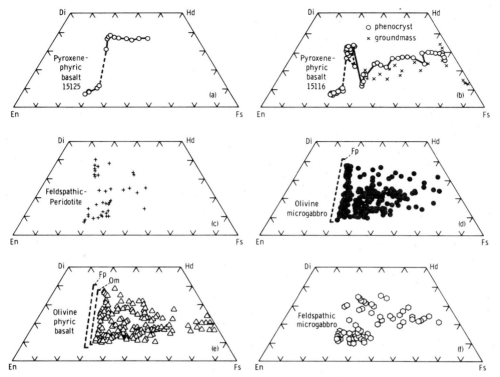

Fig. 4.36. Composition of pyroxenes from mare basalts projected onto the pyroxene quadrilateral. In (a) 15125 and (b) 15116, microprobe traverses from center to edge of phenocrysts are shown together with random groundmass analyses in 15116; the groundmass is too fine to analyze in 15125. In (c), (d), (e) and (f) the analyses were made at random in the sections. Note the high-magnesium limit to the compositional field in each case; the limit for feldspathic peridotite (c) has been shown by the dashed line marked "Fp" on the diagrams for olivine microgabbro (d) and olivine-phyric basalt (e). The limit for olivine micrograbbro is shown by the dashed line marked "Om" in (e) (from Dowty, *et al.*, 1973; *Proc. 4th Lunar Sci Conf., Suppl. 4, Vol. 1*, Pergamon).

is not always obvious optically. Zoning generally involves an increase in both Na and K adjacent to areas of late stage crystallization. The main body of the plagioclase grain is generally close to An_{90} whereas in the outer zones the composition drops below An_{80} (Dowty *et al.*, 1973). Nearly all of the feldspars are twinned with the majority having (010) of vicinal faces as composition planes (Wenk *et al.*, 1972). Simple or polysynthetic albite twins might predominate although other twin laws occur in small numbers. As might be expected, twinning of the lunar plagioclases is very similar to their terrestrial counterparts.

The olivine content of the mare basalts is variable — up to 34 per cent in some varieties and absent in others. The olivine grains are generally rich in magnesium with peaks in the range of Fo_{50} to Fo_{65} depending upon the individual lithology. In some lithologies a minor late-stage fayalitic olivine may be present (Dowty et al., 1973). Zoning is common in some lithologies and in a few samples extremely skeletal olivine phenocrysts are present.

Ilmenite is the most common opaque mineral and is particularly conspicuous in rocks returned from Mare Tranquillitatis by Apollo 11. The other opaque phases are mostly metals or sulphides as a result of the extremely reducing conditions under which the basalts crystallized.

Silica in the form of cristobalite and tridymite occurs in some basalt types but is never abundant. Tridymite laths are seen transecting the margins of pyroxene and plagioclase crystals whereas cristobalite occurs in the intergranular areas only. Tridymite thus crystallized earlier than cristobalite suggesting that the cristobalite must have been metastable (Mason, 1972).

Texturally, the lunar basalts are also extremely varied and at least at the Apollo 11 and 12 sites may be separated into three groups (1) Porphyritic basalts, (2) Ophitic basalts, and (3) Intersertal basalts (Warner, 1971). These three groups may be further subdivided into thirteen types on the basis of the nature of the matrix and phenocrysts in the porphyritic group, the plagioclase-pyroxene texture in the ophitic group and the grain size of the pyroxene-ilmenite network in the intersertal group.

The *porphyritic basalts* have pyroxene or plagioclase phenocrysts (sometimes both) in a glass matrix or in a matrix of fine-grained acicular crystals of augite plagioclase and opaques (Fig. 4.37 a). Modally these basalts are extremely variable. The *ophitic basalts* have a distinctive texture in which the mineral phases form a continuous size-graded series (Fig. 4.37b). The phases include subhedral to anhedral pyroxene, subhedral olivine, euhedral plagioclase laths or plate-like opaques and intersertal cristobalite (Warner, 1971; James and Jackson, 1970). This textural group shows little modal variation. The *intersertal basalts* are distinguished by a network of subhedral pyroxenes and skeletal or equant opaques of approximately equal size (Fig. 4.37c). Large and partially resorbed olivines are distributed throughout the network of pyroxenes and opaques. The interstitial areas are filled with euhedral plagioclase, anhedral cristobalite and glass.

Chemically the mare basalts are distinct from their terrestrial counterparts (Table 4.5). In general the iron and titanium content of the lunar lavas are higher and the sodium and potassium content is lower. Experimental data suggest that the lunar basalts were formed by local partial melting at depths of 200-400 km (Ringwood and Essene, 1970; Green et al., 1971). The genesis of the mare basalts appears more complex than current models would suggest. Consequently, the models are presently under considerable revision and this makes further discussion difficult.

Fig. 4.37. Photomicrographs of typical lunar mare basalt textural varieties. (a) Porphyritic, (b) Ophitic and (c) Intersertal. The longest dimension of each photo is 3.1 mm (NASA Photos S-70-49563, S-70-49438, S-70-48984).

TABLE 4.5

Range in composition of mare basalts and the 10-90 per cent frequency range of terrestrial continental basalts (from Taylor, 1975).

	Moon	Earth
SiO_2	37—49	44—53
TiO_2	0.3—13	0.9—3.3
Al_2O_3	7—14	13—19
FeO	18—23	7—14
MnO	0.21—0.29	0.09—0.3
MgO	6—17	4—10
CaO	8—12	8—12
Na_2O	0.1—0.5	1.8—3.8
K_2O	0.02—0.3	0.3—2.0
P_2O_5	0.03—0.18	0.04—0.6
Cr_2O_3	0.12—0.70	0.005—0.04

Eratosthenian and Copernican Systems

Crater Materials

The Imbrian System is followed by the Eratosthenian System and finally by the youngest time-stratigraphic unit, the Copernican System. By definition these units are superimposed on the Imbrian; there are, however, many unresolved problems both in separating the Eratosthenian from the Copernican and separating both from the Imbrian. The main material units associated with the Eratosthenian and Copernican Systems are ejecta blankets surrounding craters. The Eratosthenian craters are generally fresh but rayless whereas the younger Copernican craters are surrounded by bright rays. The reason for using rays as a criterion is that they appear to be among the most recent materials on the moon and are superimposed on most other material units. Copernicus, the type-crater example, has a very extensive ray system which overlies the nearly rayless crater Eratosthenes. Secondary processes, such as micrometeoroid reworking, cause the rays to darken and disappear with time. The criterion is not entirely satisfactory. If, for example, rays fade due to mixing with darker materials, the rays surrounding small craters would fade faster than those surrounding large craters. Furthermore, there are craters with dark haloes clearly superimposed on the ray and rim materials of craters with light rays. Similar craters on surfaces with a low albedo would not be identifiable as Copernican in age.

The production of glass particles by micrometeoroid reworking and in particular the production of agglutinates (complex compound glass particles) causes a lowering of the albedo of lunar soil (Adams and McCord, 1974). Agglutinates are typically darker than the parent soil due to the presence of iron and titanium ions in solution in the glass and to finely disseminated materials in the glass. It has been argued that this is the principal darkening mechanism on the lunar surface (Adams and McCord, 1974). If so, it may well explain the fading of crater rays in which case fading would be directly time dependent.

Mantling Materials .

Several types of mantling material have been observed on the moon. The oldest of these mantling materials are dark terra-mantling units which are generally darker than the mare and occur on the uplands near the mare margins (Fig. 4.38). These deposits may extend over as much as 500,000 km² (Head, 1974). Several of the areas shown in Figure 4.38 have been assigned formational names. The name Sulpicius Gallus Formation has been applied to dark mantling deposits on the southern edge of Mare Serenitatis and to similar deposits in areas adjacent to Mare Vaporum and Sinus Aesthuum and near Copernicus. Similar materials adjacent to Mare Humorum have been called the Doppelmayer Formation (Wilhelms, 1968, 1970;

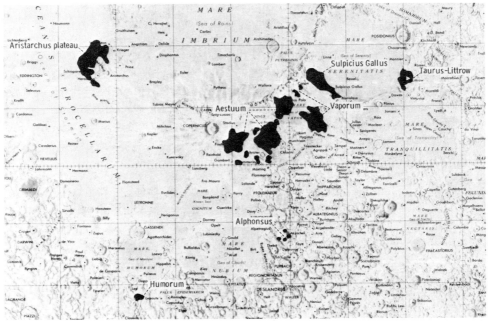

Fig. 4.38. Major lunar dark-mantle deposits. Location and extent (from Head, 1974; *Proc. 5th Lunar Sci. Conf., Suppl. 5, Vol. 1,* Pergamon).

Carr, 1966; Titley, 1967).

The surface of the dark mantle materials is generally smooth with no blocks or boulders visible. It appear to be draped over the antecedent topography and is thickest in topographic lows and may be absent entirely on some hilltops which appear as bright areas. The apparent rapid degradation of craters and the bare hilltops suggests that the dark mantle is unconsolidated.

The Sulpicius Gallus Formation has been found to be thickest along rilles, suggesting that this may be its source. The Apollo 17 mission landed in an area of dark mantling materials on the southeastern edge of Mare Serenitatis (Fig. 4.38). Initally there was confusion as to whether the dark mantle had been sampled until it was realized that orange and black glasses occuring in varying proportions in most of the soils were probably pyroclastic in origin. The black glass spheres which are relatively abundant in the soil appear to be the devitrified equivalent of the orange glasses; both probably originated in a volcanic fire fountain (Reid *et al.*, 1973; McKay and Heiken, 1973; Carter *et al.*, 1973). These spheres are spectrally the characteristic component of the dark mantle (Adams *et al.*, 1974). It therefore seems likely that the Sulpicius Gallus and Doppelmayer Formations consist of pyroclastic materials which have been reworked, at least closer to the surface, by the normal soil forming processes. Similar green glass spheres from Apollo 15 soils suggest that these pyroclastic deposits may be much more wide spread than even the distinctive dark mantle would suggest.

Orange and red materials occur in the Sulpicius Gallus Formation on the southeastern rim of the Serenitatis Basin (Lucchitta and Schmitt, 1974). The orange and red materials occur as patches, haloes or rays around small craters between 50 and 250 m in diameter. In larger craters the orange and red materials are exposed in layers underlying the dark mantle deposits to a depth of about 50 m.

The orange soil returned by Apollo 17 consists largely of dark-orange-brown glass particles with a mean size of 44 μm (4.51ϕ) (Lindsay, 1974) which is finer than the typical texturally-mature lunar soil (58 μm or 4.1ϕ). The glass particles are not particularly well sorted (σ = 1.39ϕ) and are comparable with normal texturally mature soils. The addition of pyroclastic materials to the lunar soil has minimal effect on its evolution unless they are added very early in soil development and in large volumes (Lindsay, 1974).

The dark orange-brown glass particles, which are the most abundant particle type, are generally non-vesicular, transparent, homogeneous, rounded forms such as beads, wafers and teardrops (Carter *et al.*, 1973). Two types of opaque particles are also present, one with a dust coated surface the other consisting of alternating layers of glass and olivine (Fo_{67-84}). The second glass type may also contain a feathery Ti-rich oxide phase (armalcolite) and a chromium-ulvöspinel. The average composition of these complex layered particles is the same as that of the homogeneous glasses (Table 4.6).

Compositionally the orange glasses are also very similar to the associated mare basalts from the Apollo 17 site suggesting a close genetic association. There are minor chemical differences, for example they are somewhat richer in MgO and volatile elements than the associated basalts (Table 4.6).

Carr (1966), who first defined the Sulpicius Gallus Formation, considered the unit to be of Imbrian and Eratosthenian age. It clearly overlies the Fra Mauro Formation and in some areas appears to overlie mare materials. In other areas the dark mantle appears to be embayed by mare basalts. Radiometric ages for the orange glasses range from 3.5 to 3.83 aeons (Schaeffer and Husain, 1973; Tatsumoto et al., 1973). In comparison the mare basalts from the Apollo 17 site have ages ranging form 3.74 to 3.83 (Table 4.3). It is thus apparent that the dark mantle and the mare basalts are closely related and that their formation began in Imbrian time. How far the pyroclastic activity extended into Eratosthenian time is not certain. Head (1974) has suggested that the pyroclastic materials are associated with an early phase of basalt flooding in which case they may all be Imbrian in age. However, photogeologic information suggests that some dark-mantle materials are Eratosthenian and it seems more reasonable to suggest that the age of

TABLE 4.6

Chemical composition of glass particles from the Apollo 17 orange soil (from Carter et al., 1973).

Wt. %	Orange Glass	"Barred" Particle			Dust Coated
		Glass	Olivine	Ulvöspinel	
SiO_2	38.0	39.6	35.8	0.7	45.2
TiO_2	8.87	11.8	1.4	29.4	0.49
Al_2O_3	5.51	7.58	0.3	3.3	23.4
Cr_2O_3	0.70	0.55	0.6	10.6	0.17
FeO	22.4	21.4	27.1	47.8	6.77
MgO	14.5	6.73	34.6	3.8	9.25
CaO	6.99	11.1	0.5	0.5	14.0
Na_2O	0.39	0.48	0.0	0.1	0.30
K_2O	0.06	0.06	0.0	0.0	0.06
TOTAL	97.42	99.27	100.3	96.2	99.71

the dark-mantle materials probably varies considerably from one locality to the next.

There are some extensive dark mantling units, similar to the Sulpicius Gallus and Dopplemayer Formations, that are assigned an Eratosthenian and Copernican age. The units appear thin and add little to the topography of the antecedent terrain. One of these units had been formerly defined as the Cavalerius Formation (McCauley, 1967). It is superimposed on the rim material of the post-mare crater Cavalerius, on the surrounding mare surface, and on Copernican rays. Consequently, the unit has been assigned a Copernican age.

The Reiner Gamma Formation (7° N, 59° W) is worthy of special mention in that it is a mantling unit that has a higher albedo than the underlying unit (McCauley, 1967a; Wilhelms, 1970). As with the dark mantling materials it is thought to be volcanic and has been assigned an Eratosthenian and Copernican age. Clearly volcanism continued in a small way at least for a considerable period of time on the lunar surface after completion of the mare filling.

Formations with Morphologic Expression

A number of units of probable constructional origin have been identified on the moon with ages ranging from Imbrian to Copernican. The Harbinger Formation, which occurs to the east of the Aristarchus Plateau (Fig. 4.23) appears to consist of volcanic domes, cones and craters although some of its relief may be antecedent Fra Mauro topography. This formation is transected by numerous sinuous rilles, the high ends of which begin in craters while the low ends terminate at the level of the mare. This adds further credence to the volcanic origin. The Harbinger Formation may interfinger with the mare materials locally although in other areas it appear to embay mare material. Consequently, it has been designated Imbrian or Eratosthenian in age. The activity of this volcanic complex appear to have moved westward at a later time and resulted in the development of at least part of the Aristarchus Plateau. The Vallis Schröteri Formation forms part of the Aristarchus Plateau (Fig. 4.23) and has been assigned an Eratosthenian or Copernican age. As well as thin blanketing materials this formation includes plains-forming materials, low rimmed craters and crater domes. Like the Harbinger Formation the Vallis Schröteri Formation is transected by sinuous rilles which terminate in the mare. These two formations probably represent continued volcanism the focal point of which simply change with time.

A further complex of younger domes occurs in the northwest quadrant of the moon. The Marius Group has been assigned an Eratosthenian age (McCauley, 1967). The region consists of undulating plateau forming materials and two types of domes of possible volcanic origin (Fig. 4.39). One dome type is low and convex in profile, the other has steeper slopes with a concave profile.

Post-Imbrian Stratigraphy and the Lunar Soil

Since the lunar stratigraphy was initially established using earth based telescope photography and later orbital photography some of the subtlety of

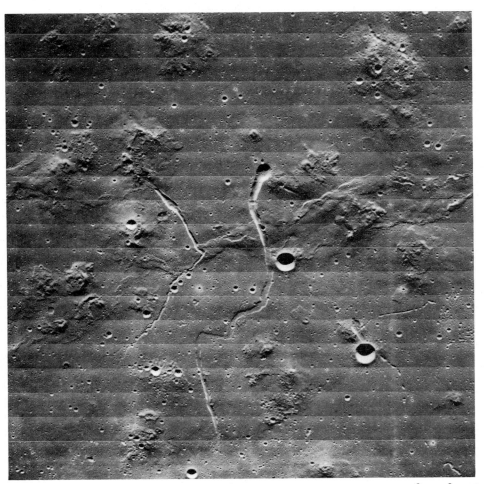

Fig. 4.39. The Marius Hills northwest of the crater Marius in Oceanus Procellarum at 14° N, 15° W. The numerous cones and domes are believed to be volcanic. There are two types of domes; gently rising smooth low domes and rugged heavily cratered steep-sided domes. Wrinkle ridges transect the area and are themselves transected by sinuous rilles. The two larger craters are 10 km in diameter (NASA Orbiter Photo V-M214).

the record must inevitably be lost. This is particularly true in the time period following the extrusion of the mare basalts. The ejecta blankets upon which the post-Imbrian stratigraphy was established are relatively small when compared to the multiringed basins. Consequently, it is difficult to establish an extensive stratigraphic history for this time period. However, all through the post-mare period a soil stratigraphy was evolving on the lunar surface. The Apollo missions allowed a detailed sampling of this thin but extensive sedimentary body which will ultimately allow a much more complete study of the late stage history of the moon. Because of the importance of the soil a separate chapter, Chapter 6 is devoted entirely to it.

References

Adams, J. B. and McCord, T. B., 1974. Vitrification darkening in the lunar highlands and identification of Descartes material at the Apollo 16 site. *Proc. Fourth Lunar Sci. Conf., Suppl. 4, Geochim. Cosmochim. Acta, 1*: 163-177.

Adams, J. B., Pieters, C. and McCord, T. B., 1974. Orange glass: evidence for regional deposits of pyroclastic origin on the moon. *Proc. Fifth Lunar Sci. Conf., Suppl. 5, Geochim. Cosmochim. Acta, 1*: 171-186.

Arkani-Hamed, J., 1974. Effects of a giant impact on the thermal evolution of the moon. *The Moon, 9*: 183-209.

Baldwin, R. B., 1949. *The Face of the Moon.* Univ. Chicago Press, Chicago, 239 pp.

Baldwin, R. B., 1963. *The Measure of the Moon.* Univ. Chicago Press, Chicago, 488 pp.

Baldwin, R. B., 1972. The Tsunami Model of the origin of ring structures concentric with large lunar craters. *Phys. Earth Planet. Interiors, 6*: 327-339.

Barrell, J., 1927. On continental fragmentation, and the geology of the moon's surficial features. *Amer. Jour. Sci., 5th ser., 13*: 283-314.

Bence. A. E. and Papike, J. J., 1972. Pyroxenes as recorders of lunar basalt petrogenesis: chemical trends due to crystal-liquid interaction. *Proc. Third Lunar Sci. Conf., Suppl. 3, Geochim. Cosmochim. Acta, 1*: 431-469.

Bence, A. E., Papike, J. J., Sueno, S., and Delani, J. W., 1973. Pyroxene poikilobastic rocks from the lunar highlands. *Proc. Fourth Lunar Sci. Conf., Suppl. 4, Geochim. Cosmoshim. Acta, 1*: 597-611.

Brown, W. E., Adams, G. F., Eggelton, R. E., Jackson, P., Jordan, R., Kobrick, M., Peeples, W. J., Phillips, R. J., Porcello, L. J., Schaber, G., Sill, W. R., Thompson, T. W., Ward, S. H. and Zelenka, J. S., 1974. Elevation profiles of the moon. *Proc. Fifth Lunar Sci. Conf., Suppl. 5, Geochim. Cosmochim. Acta, 3*: 3037-3048.

Bryan, W. B., 1973. Wrinkle-ridges as deformed surface crust on ponded mare lava. *Proc. Fourth Lunar Sci. Conf., Suppl. 4, Geochim. Cosmochim. Acta, 1*: 93-106.

Carr, M. H., 1966. Geologic map of the Mare Serenitatis region of the moon. *U. S., Geol. Survey Misc. Geol. Inv. Map.* I-489.

Carr, M. H., 1974. The role of lava erosion in the formation of lunar rilles and Martian channels. *Icarus, 22*: 1-23.

Carter, J. L., Taylor, H. C., and Padovani, E., 1973. Morphology and chemistry of particles form Apollo 17 soils 74220, 74241, 75081. *EOS, 54*: 582-584.

Chao, E. C. T., Soderblom, L. A., Boyce, J. M., Wilhelms, D. E., and Hodges, C. A., 1973. Lunar light plains deposits (Cayley Formation) — reinterpretation of origin. *Lunar Science—V,* The Lunar Science Inst., Houston, Texas, 127-128.

Dence, M. R., 1974. The Imbrium basin and its ejecta. *Lunar Science—V,* The Lunar Science Inst., Houston, Texas, 165-167.

Dowty, E., Prinz, M. and Keil, K., 1973. Composition, minerology, and petrology of 28 mare basalts from Apollo 15 rake samples. *Proc. Fourth Lunar Sci. Conf., Suppl. 4, Geochim. Cosmochim. Acta, 1*: 423-444.

Eberhardt, P., Geiss, J., Grögler, N. and Stettler, A., 1973. How old is the crater Copernicus? *The Moon, 8*: 104-114.

Eggleton, R. E., 1963. Thickness of the Apenninian Series in the Lansberg region of the moon. *Astrogeol. Studies Ann. Prog. Rept., August, 1961—August, 1962: U. S. Geol. Survey open-file rept.,* 19-31.

Eggleton, R. E., 1964. Preliminary geology of the Riphaeus quadrangle of the moon and a definition of the Fra Mauro Formation. *Astrogeol. Studies Ann. Rept., August, 1962—July, 1963, pt. A: U. S. Geol. Survey open-file report,* 46-63.

Eggelton, R. E., 1965. Geologic map of the Riphaeus Mountains region of the moon. *U. S. Geol. Survey Misc. Geol. Inv. Map.* I-458.

Eggleton, R. E. and Offield, T. W., 1970. Geologic map of the Fra Mauro region of the moon. *U. S. Geol. Survey Misc. Geol. Inv. Map.* I-708.

Eggleton, R. E. and Schaber, G. G., 1972. Cayley Formation interpreted as basin ejecta. *NASA SP-315:* 29-5 to 29-16.

El-Baz, F., 1973. Al-Khwarizmi: a new-found basin on the lunar farside. *Science, 180:* 1173-1176.

Ferrari, A. J., 1975. Lunar gravity: the first far-side map. *Science, 188:* 1297-1299.

Fielder, G., 1963. Nature of the lunar maria. *Nature, 198:* 1256-1260.

Gilbert, G. K., 1893. The Moon's face, a study of the origin of its features. *Phil. Soc. Washington Bull., 12:* 241-292.

Gornitz, V., 1973. The origin of sinuous rilles. *The Moon, 6:* 337-356.

Green, D. H., Ringwood, A. E., Ware, N. G., Hibberson, W. O., Major, A. and Kiss, E., 1971. Experimental petrology and petrogenesis of Apollo 12 basalts. *Proc. Second Lunar Sci. Conf., Suppl. 2, Geochim. Cosmochim. Acta, 1:* 601-615.

Hackmann, R. J., 1962. Geologic map of the Kepler region of the moon. *U. S. Geol. Survey Misc. Geol. Inv. Map.* I-355.

Hackmann, R. J., 1964. Stratigraphy and structure of the Montes Apenninus quadrangle of the moon. *Astrogeol. Studies Ann Prog. Rept., August 1962—July 1963, Pt. A: U. S. Geol. Survey open-file report,* 1-8.

Hackmann, R. J., 1966. Geologic map of the Montes Apenninus region of the moon. *U. S. Geol. Survey Misc. Geol. Inv. Map.* I-463.

Hartmann, W. K., 1972. Interplanet variations in scale of crater morphology Earth, Mars, Moon. *Icarus, 17:* 707-713.

Hartmann, W. K. and Kuiper, G. P., 1962. Concentric structures surrounding lunar basins. *Arizona Univ. Lunar Planet. Lab. Comm. 1:* 51-66.

Hartmann, W. K. and Wood, C. A., 1971. Moon: origin and evolution of multiringed basins. *The Moon, 3:* 3-78.

Hartmann, W. K. and Yale, F. G., 1969. Mare Orientale and its intriguing basin. *Sky and Telescope, 37:* 4-7.

Head, J. W., 1974. Lunar dark-mantle deposits: possible clues to the distribution of early mare deposits. *Proc. Fifth Lunar Sci. Conf., Suppl 5, Geochim. Cosmochim. Acta, 1:* 207-222.

Hodges, C. A., Muehlberger, W. R. and Ulrich, G. E., 1973. Geologic setting of Apollo 16. *Proc. Fourth Lunar Sci. Conf., Suppl. 4, Geochim. Cosmochim. Acta, 1:* 1-25.

Hörz, F., Oberbeck, V. R., and Morrison, R. H., 1974. Remote sensing of the Cayley plains and Imbrium basin deposits. *Lunar Science—V,* The Lunar Science Inst., Houston, Texas, 357-359.

Howard, K. A., Wilhelms, D. E., and Scott, D. H., 1974. Lunar basin formation and highland stratigraphy. *Rev. Geophys. Space Phys., 12:* 309-327.

Howard, K. A., Head, J. W., and Swann, G. A., 1972. Geology of Hadley Rille. *Proc. Third Lunar Sci. Conf., Suppl. 3, Geochim. Cosmochim. Acta, 1:* 1-14.

Huneke, J. C., Jessberger, E. K., Podesek, F. K., and Wasserburg, G. J., 1973. $^{40}Ar/^{39}Ar$ measurements in Apollo 16 and 17 samples and the chronology of metamorphic and volcanic activity in the Taurus-Littrow region. *Proc. Fourth Lunar Sci. Conf., Suppl. 4, Geochim. Cosmochim. Acta, 2:* 1725-1756.

Husain, L. and Schaeffer, O. A., 1975. Lunar evolution: the first 600 million years. *Geophys. Res. Lett., 2:* 29-32.

James, O. B., and Jackson, E. D., 1970. Petrology of the Apollo 11 ilmenite basalts. *Jour. Geophys. Res., 75:* 5793-5824.

Kaula, W. M., Schubert, G., Lingenfelter, R. E., Sjögren, W. L., and Wollenhaupt, W. K., 1972. Analysis and interpretation of lunar laser altimetry. *Proc. Third Lunar Sci. Conf., Suppl. 3, Geochim. Cosmochim. Acta, 3:* 2189-2204.

Kaula, W. M., Schubert, G., Lingenfelter, R. E., Sjögren, W. L., and Wollenhaupt, W. K., 1973. Lunar topography from Apollo 15 and 16 laser altimetry. *Proc. Fourth Lunar Sci. Conf., Suppl. 4, Geochim. Cosmochim. Acta, 3:* 2811-2819.

Kaula, W. M., Schubert, G., Lingenfelter, R. E., Sjögren, W. L., and Wollenhaupt, W. R., 1974. Apollo laser altimetry and inferences as to lunar structure. *Proc. Fifth Lunar Sci. Conf., Suppl. 5, Geochim. Cosmochim. Acta, 3:* 3049-3058.

Khabakov, A. V., 1962. Characteristic features of the relief of the moon – basic problems of the genesis and sequence of development of lunar formations. In: A. V. Markov (Editor), *The Moon, a Russian view*, Univ. Chicago Press, Chicago, 247-303.

Kovach, R. L., Watkins, J. S. and Landers, T., 1971. Active seismic experiment. In: *Apollo 14 Preliminary Science Report.* NASA, SP-272, 163-174.

Kuiper, G., 1954. On the origin of the lunar surface features. *Proc. Nat. Acad. Sci., 40:* 1096-1112.

Latham, G., Dorman, J., Duennebier, F., Ewing, M., Lammlein, D. and Nakamura, Y., 1974. Moonquakes, meteoroids, and the state of the lunar interior. *Proc. Fourth Lunar Sci. Conf., Suppl. 4, Geochim. Cosmochim. Acta, 3:* 2515-2527.

Lindsay, J. F., 1974. A general model for the textural evolution of the lunar soil. *Proc. Fifth Lunar Sci. Conf., Suppl. 5, Geochim. Cosmochim. Acta, 1:* 861-878.

Lingenfelter, R. E., Peale, S. J. and Schubert, G., 1968. Lunar rivers. *Science, 161:* 266-269.

Lucchitta, B. K. and Schmitt, H. H., 1974. Orange material in the Sulpicius Gallus Formation at the southwestern edge of Mare Serenitatis. *Proc. Fifth Lunar Sci. Conf., Suppl. 5, Geochim. Cosmochim. Acta, 1:* 223-234.

McCauley, J. F., 1964. The stratigraphy of the Mare Orientale region of the moon. In: *Astrogeol. Studies Ann. Prog. Rept., August 1962 – July 1963, Pt. A: U. S. Geol. Survey open-file report,* 86-98.

McCauley, J. F., 1967. Geologic map of the Hevelius region of the moon. *U. S. Geol. Survey Inv. Map.* I-491.

McCauley, J. F., 1968. Preliminary geologic map of the Orientale basin region. *U. S. Geol. Survey Interagency Rept., Astrogeol., 7:* 32-33.

McCauley, J. F., 1969. Geologic map of the Alphonsus GA region of the moon. *U. S. Geol. Survey Misc. Inv. Map.* I-586.

McGetchin, T. R., Settle, M. and Head, J. W., 1974. Radial thickness variation in impact crater ejecta: implications for lunar basin deposits. *Earth Planet. Sci. Lett., 20:* 226-236.

McGill, G. E., 1971. Attitude of fractures bonding straight and arcuate lunar rilles. *Icarus, 14:* 53-58.

McKay, D. S. and Heiken, G. H., 1973. Petrography and scanning electron microscope study of Apollo 17 orange and black glass. *EOS, 54:* 599-600.

Marshall, C. H., 1961. Thickness of the Procellarian System, Letronne region of the moon. *U. S. Geol. Survey Prof. Pap. 424D:* 208-211.

Marshall, C. H., 1963. Geologic map of the Letronne region of the Moon. *U. S. Geol. Survey Misc. Geol. Inv. Map.* I-385.

Mason, B. 1972. Lunar tridymite and cristobalite. *Amer. Mineral., 57:* 1530-1535.

Milton, D. J., 1968. Geologic map of the Theophilus quadrangle of the moon. *U. S. Geol. Survey Misc. Geol. Inv. Map.* I-546.

Morris, E. C. and Wilhelms, D. E., 1967. Geologic map of the Julius Caesar quadrangle of the moon. *U. S. Geol. Survey Misc. Inv. Map.* I-510.

Muller, P. M. and Sjögren, W. L., 1968. Mascons: Lunar mass concentrations. *Science, 161:* 680-684.

Mutch, T. A., 1970. *Geology of the Moon.* Princeton Univ. Press, N.J., 324 pp.

Neukem, G. and König, B., 1975. A study of lunar impact crater size distribution. *The Moon, 12:* 201-229.

Numes, P. D., Tatsumoto, M., Knight, R. J., Unruh, D. M. and Doe, B. R., 1973. U-Th-Pb systematics of some Apollo 16 lunar samples. *Proc. Fourth Lunar Sci. Conf., Suppl. 4, Geochim. Cosmochim. Acta, 2:* 1797-1822.

Oberbeck, V. R., Aoyagi, M., Greely, R. and Lovas, M., 1972. Planimetric shapes of lunar rilles. In: *Apollo 16 Preliminary Science Report.* NASA SP-315, 29-80 to 29-88.

Oberbeck, V. R., Morrison, R. H., Hörz, F., Quaide, W. L. and Gault, D. E., 1974. Smooth plains and continuous deposits of craters and basins. *Proc. Fifth Lunar Sci. Conf., Suppl. 5, Geochim. Cosmochim. Acta, 1:* 111-136.

Papanastassiou, D. A. and Wasserburg, G. J., 1971. Rb-Sr ages of igneous rocks from the Apollo 14 mission and the age of the Fra Mauro Formation. *Earth Planet. Sci. Lett., 12:* 36-48.

Papanastassiou, D. A. and Wasserburg, G. J., 1972. Rb-Sr ages and initial strontium in basalts from Apollo 15. *Earth Planet. Sci. Lett. 17*: 324-337.

Phillips, R. J. and Saunders, R. S., 1974. Interpretation of gravity anomalies in the irregular maria. *Lunar Science—V*, The Lunar Science Inst., Houston, Texas, 596-597.

Reid, A. M., Ridley, W. I., Donaldson, C. and Brown, R. W., 1973. Glass compositions in the orange and gray soils from Shorty Crater, Apollo 17. *EOS 54*: 607-609.

Ringwood, A. E. and Essene, E. J., 1970. Petrogenesis of Apollo 11 basalts, internal constitution, and origin of the moon. *Proc. Apollo 11 Conf., Suppl. 1, Geochim. Cosmochim. Acta, 1*: 796-799.

Schaber, G. G., 1973. Lava flows in Mare Imbrium: geologic evaluation from Apollo orbital photography. *Proc. Fourth Lunar Sci. Conf., Suppl. 4, Geochim. Cosmochim. Acta, 1*: 73-92.

Schaeffer, O. A. and Husain, L., 1973. Isotopic ages of Apollo 17 lunar material. *EOS, 54*: 614.

Schaeffer, O. A. and Husain, L., 1974. Chronology of lunar basin formation. *Proc. Fifth Lunar Sci. Conf., Suppl. 5, Geochim. Cosmochim. Acta, 2*: 1541-1555.

Schubert, G., Lingenfelter, R. E. and Peale, S. J., 1970. The morphology, distribution and origin of lunar sinuous rilles. *Rev. Geophys. Space Phys., 8*: 199-225.

Schumm, S. A., 1970. Experimental studies on the formation of lunar surface features by fluidization. *Geol. Soc. Amer. Bull., 81*: 2539-2552.

Schubert, G., Lingenfelter, R. E., and Kaula, W. M., 1974. Lunar sinuous rille elevation profiles. *Lunar Science—V*, The Lunar Science Inst., Houston, Texas, 675-677.

Shoemaker, E. M., 1962. Interpretation of lunar craters. In: Z. Kopal (Editor), *Physics and Astronomy of the Moon*. Academic Press, N. Y., 283-359.

Shoemaker, E. M. and Hackmann, R. J., 1962. Stratigraphic basis for a lunar time scale. In: Z. Kopal and Z. K. Mikhailov (Editors), *The Moon* — Symp. 14, Internat. Astron. Union, Leningrad, 1960. Academic Press, London.

Shoemaker, E. M., Hackmann, R. J., Eggleton, R. E. and Marshall, C. H., 1962. Lunar stratigraphic nomenclature. *Astrogeol. Studies Semiann. Prog. Rept., February 1961-August 1961*: U. S. *Geol. Survey open-file report, 114-116.*

Spurr, J. E., 1944-1949. *Geology applied to selenology.* Vols. 1 and 2, Science Press, Lancaster, Pa., Vols. 3 and 4, Rumford Press, Concord, N.H.

Stuart-Alexander, D. and Howard, K. A., 1970. Lunar maria and circular basins — a review. *Icarus, 12*: 440-456.

Tatsumoto, M., 1970. Age of the moon: an isotopic study of U-Th-Pb systematics of Apollo 11 lunar samples, II. *Proc. Second Lunar Sci. Conf., Suppl. 2, Geochim. Cosmochim. Acta, 2*: 1595-1612.

Tatsumoto, M., Nunes, P. D., Knight, R. J., Hedge, C. E. and Unruh, D. M., 1973. U-Th-Pb, Rb-Sr, and K measurements of two Apollo 17 samples. *EOS, 54*: 614-615.

Taylor, S. R., 1975. *Lunar science: a post-Apollo view.* Pergamon Press. N. Y., 372 pp.

Tera, F. and Wasserburg, G. J., 1974. U-Th-Pb systematics on lunar rocks and inferences about lunar evolution and the age of the moon. *Proc. Fifth Lunar Sci. Conf., Suppl. 5, Geochim. Cosmochim. Acta, 2*: 1571-1599.

Tera, F., Papanastassiou, D. and Wasserburg, G. J., 1973. A lunar cataclysm at ⁓3.95 AE and the structure of the lunar crust. *Lunar Science—IV*, The Lunar Science Inst., Houston, Texas, 723-725.

Tera, F., Papanastassiou, D. A., and Wasserburg, G. J., 1974a. Isotopic evidence for a terminal lunar cataclysm. *Earth Planet. Sci. Lett., 22*: 1-21.

Tera, F., Papanastassiou, D. A., and Wasserburg, G. J., 1974b. The lunar time scale and a summary of isotopic evidence for a terminal cataclysm. *Lunar Science—V*, The Lunar Science Inst., Houston, Texas, 792-794.

Titley, S. R., 1967. Geologic map of the Mare Humorum region of the moon. *U. S. Geol. Survey Misc. Geol. Inv. Map.* I-495.

van Dorn, W. G., 1968. Tsunamis on the moon? *Nature, 220*: 1102-1107.

Walker, D., Longhi, J. and Hays, J. F., 1973. Petrology of Apollo 16 meta-igneous rocks. *Lunar Science—IV*, The Lunar Science Inst., Houston, Texas, 752-754.

Warner, J. L., 1971. Lunar crystalline rocks: petrology and geology. *Proc. Second Lunar Sci. Conf.,* *Suppl. 2, Geochim. Cosmochim. Acta, 1:* 469-480.

Warner, J. L., Simonds, C. H., and Phinney, W. C., 1973. Apollo 16 rocks: classification and petrogenetic model. *Proc. Fourth Lunar Sci. Conf., Suppl. 4, Geochim. Cosmochim. Acta, 1:* 481-504.

Wasserburg, G. J., Huneke, J. C., Papanastassiou, D. A., Rajan, R. S. and Tera, F., 1974. A summary of the lunar time scale and implications for the chronology of the solar system. *The Moon, 11:* 408-409.

Wenk, E., Glauser, A., Schwander, H. and Trommsdorff, V., 1972. Twin laws, optic orientation, and composition of plagioclases from rocks 12051, 14053, and 14310. *Proc. Third Lunar Sci. Conf., Suppl. 3, Geochim. Cosmochim. Acta, 1:* 581-589.

Wilhelms, D. E., 1965. Fra Mauro and Cayley Formations in the Mare Vaporum and Julius Caesar quadrangles. *Astrogeol. Studies Ann. Prog. Rept. July 1964-July 1965, Pt. A: U. S. Geol. Survey open-file report,* 13-28.

Wilhelms, D. E., 1968. Geologic map of the Mare Vaporum quadrangle of the moon. *U. S. Geol. Survey Misc. Geol. Inv. Map* I-548.

Wilhelms, D. E., 1970. Summary of lunar stratigraphy — telescopic observations. *U. S. Geol. Survey Prof. Pap. 599-F,* 47 pp.

Wilshire, H. G., Stuart-Alexander, D. E. and Jackson, E. D., 1973. Apollo 16 rocks, petrology and classification. *Jour. Geophys. Res., 78:* 2379-2392.

Wollenhaupt, W. R., and Sjogren, W. L., 1972. Apollo 16 laser altimeter. In: *Apollo 16 Preliminary Science Report.* NASA SP-315: 30-1 to 30-5.

Young, R. A., Brennan, W. J., Wolfe, R. W. and Nichols, D. J., 1973. Analysis of lunar mare from Apollo photography. *Proc. Fourth Lunar Sci. Conf., Suppl. 4, Geochim. Cosmochim. Acta, 1:* 57-71.

CHAPTER 5

Lithology and Depositional History of Major Lunar Material Units

Introduction

The Apollo landing sites were all selected to investigate important stratigraphic problems and to sample major material units. Apollos 11, 12 and 15 landed upon and sampled the mare surface. Apollo 14 landed on the hummocky Fra Mauro terrain to the south of Mare Imbrium and sampled the Fra Mauro Formation. Apollo 15 investigated the Apennine Front close to the source of the Fra Mauro Formation on the margin of Mare Imbrium. Apollos 16 and 17 were intended to investigate specific problems such as late stage volcanism, and landed to sample the Cayley Formation and the dark-mantle respectively. The result is that we now have some direct knowledge of the lithology of these major lunar material units. The volcanic units which are essentially outside the scope of this book are described briefly in Chapter 4. In this chapter the Fra Mauro Formation and the crystalline breccias returned by Apollo 16 will be treated; the first as being representative of the large scale material units which surround the major multi-ringed basins, the second as probably representative of the Cayley Formation — the prominent plains-forming lunar material unit. The stratigraphic relationships of these units are discussed in detail in Chapter 4.

Lithology of the Fra Mauro Formation

Apollo 14 landed west of the center of the Fra Mauro upland approximately 550 km south of the rim of the Imbrium Basin (Fig. 5.1) and close to the type area of the Fra Mauro Formation (3°40' S, 17°27' W) (Eggleton, 1963, 1964; Wilhelms, 1970). In the region of the landing site the formation is up to 500 m thick (Eggleton, 1963). The objective of the mission was to sample the Fra Mauro Formation immediately west of Cone Crater, a young blocky crater superimposed on a ridge of the formation. The crater was presumed to have excavated the Fra Mauro Formation from beneath the soil layer to a depth of 75 m. From cratering-mechanics studies it was predicted that an inverted stratigraphic sequence would be preserved

Explanation

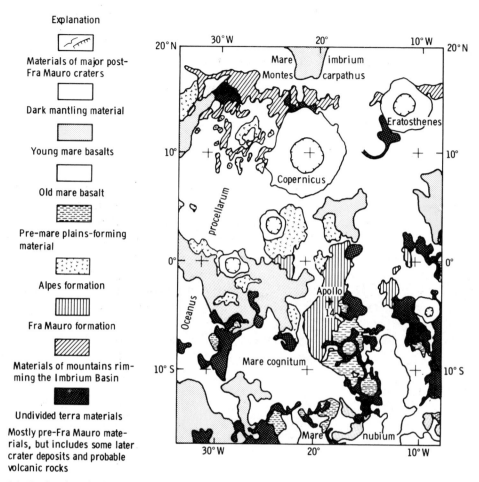

Fig. 5.1. Regional geologic map of the area surrounding the Apollo 14 landing site (from Wilshire and Jackson, 1972).

in the excavated materials and that materials from shallow depth in the crater would be ejected further from the crater (Chapter 3). Thus by traversing radially from the crater a representative sampling of the upper 75 m of the Fra Mauro Formation could be obtained in a reasonable stratigraphic context.

Megascopic Features of the Fra Mauro Lithology

The area around Cone Crater is strewn with boulders with maximum diameters of 2 m at 1000 m from the crater rim to 15 m at the rim. Three major features are visible in these large boulders: (1) a clastic texture, (2) stratification and (3) jointing or fracturing (Fig. 5.2). These features show up particularly well in large light-colored boulders near the rim of Cone Crater (Sutton *et al.*, 1972).

Both dark and light clasts are visible in the larger boulders. Most clasts have a diameter of less than 10 cm but a few reach 1.5 m. The larger clasts all appear to be reincorporated fragments of older breccias. Both stratification and jointing (or fracturing) are visible in the larger boulders. Micrometeoroid erosion has etched the boulders along such lines of weakness, producing a weathered appearance similar to that seen in terrestrial boulders (Sutton *et al.*, 1973). Some of the layering is highly irregular and contorted whereas other layers have planar contacts. Light-colored layers in the large boulders at the Cone Crater rim are as much as 1 m thick. Other boulders that are essentially monotoned have layers defined by changes in clast size (Wilshire and Jackson, 1972).

The fractures usually occur in multiple sets of generally planar, parallel

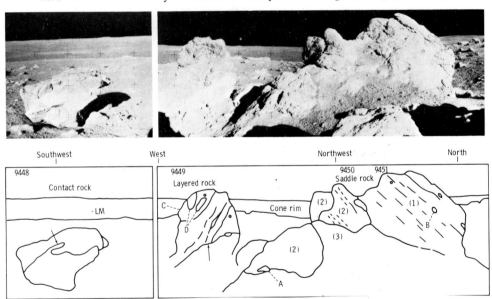

Fig. 5.2. "White Rock" group on the rim of Cone Crater. Contact Rock is about 3 m long; Layered Rock is about 3 m wide, and Saddle Rock is about 5 m long. The clasts and layers are prominent features. Most of the clasts are 10 cm or smaller, a few about 25 cm (A and B) and one (C) atop Layered Rock is a 1½ m long compound clast that contains at least two dark clasts about ½ m long at D. The dark and light layers that show in the rocks range in thickness from about 10-15 cm (dark layers in Layered Rock) to about 2 m (unit 1 in Saddle Rock). The structures expressed by these layers are a tight fold in Contact Rock, a truncated fold in Layered Rock, and in Saddle Rock, a possible fault where units 1 and 2 are in contact with unit 3. The White Rocks also contain evidence of jointing and a pervasive fracturing that probably was generated during of before the Cone Crater event (NASA Photos AS14-68-9448 to AS14-68-9451) (from Sutton *et al.*, 1972).

joints spaced from a few millimeters to a few centimeters apart. Rarely a weak lineation or foliation is produced by alignment of clasts. In general the larger scale sedimentary features visible in the boulders surrounding Cone Crater are comparable with ejecta materials surrounding Meteor Crater, Arizona (Sutton *et al.*, 1972) and terrestrial volcanic ash flow and base surge deposits.

Evidence of very large scale layers in the Fra Mauro Formation is limited due to the nature of the terrain at the Apollo 14 site. Breccias in which dark clasts predominate appear to come from deeper in the formation than those with lighter clasts (Sutton *et al.*, 1972) suggesting that large scale layering may be present. Furthermore, large scale layering was observed in the mountain faces at the Apollo 15 site on the margin of Mare Imbrium (Wolfe and Bailey, 1972) (Fig. 5.3). These layered sequences may have been formed by the impact which produced the Serenitatis Basin (Taylor, 1975). If so, such large scale layering may be a feature of the ejecta units which surround the mare basins.

Classification of Fra Mauro Formation Lithologies

Wilshire and Jackson (1972) and Chao *et al.* (1972) have both attempted to classify the Apollo 14 clastic rocks using a reasonably nongenetic framework. Both classification systems depend upon the nature of the larger clasts and the coherence of the matrix. Either system provides

Fig. 5.3. Large scale layering in Silver Spur which forms part of the Apennine Mountains to the southeast from the Apollo 15 landing site. The bluff is approximately 800 m high and was photographed from a distance of approximately 20 km (NASA Photo AS15-84-11250).

an adequate means of classification. However, in the following description
the Wilshire-Jackson classification is used because it is more independent of
genesis and more readily visualized in terms of classification of terrestrial
sedimentary rocks.

Among the rock samples returned from the Apollo 14 site Wilshire and
Jackson (1972) recognized four groupings which they designate F_1 through
F_4. All four groups are polymict and contain clasts which range from angular
to subround. The groups are isolated on the basis of the relative proportions
of light and dark clasts and on the coherence of the matrix. For purposes of
the classification clasts are defined as being larger than 1 mm.

Group F_1: This group has leucocratic lithic clasts in excess of
mesocratic and melanocratic clasts and has friable matrices (Fig. 5.4). Most
of these rocks contain more than 90 per cent leucocratic clasts. The clasts
are mainly lithic crystalline very fine-grained feldspar-rich rocks. Angular
glass fragments and glass spheres or fragments of glass sphere are common.
These rocks are fine grained and most of them appear to be soil breccias with
a light-grey matrix. Soil breccias are discussed in detail in Chapter 6 but are
treated here briefly for completeness and to allow direct comparison with
the crystalline breccias.

Group F_2: These rocks contain more leucocratic clasts than mesocratic
and melanocratic clasts but have moderately coherent to coherent matrices
(Fig. 5.4). F_2 rocks also contain a greater abundance of larger clasts than F_1.
However, some basaltic rock clasts are present. Some melanocratic clasts

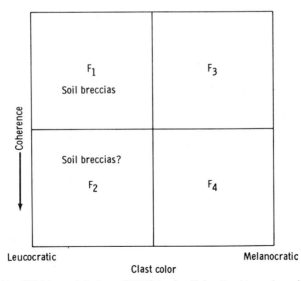

Fig. 5.4. Scheme used by Wilshire and Jackson (1972) in classifying Fra Mauro breccias.

similar to F_3 and F_4 clasts are present but not abundant. Fragments of a fine-grained olivine-pyroxene rock are also present. The color of the martix varies from light to dark grey.

Group F_3 and F_4: In these two groups the melanocratic fragments exceed the leucocratic and mesocratic fragments in numbers (Fig. 5.4). They are separated by matrix coherence; F_3 rocks being relatively friable and F_4 rocks are coherent to moderately coherent. Most of the clasts in these two groups are dark grey to black aphanitic rocks, many with metamorphic textures. Some basaltic clasts are present and clasts of the olivine-pyroxene rock are even more abundant than in group F_2. The matrix color varies from light to medium grey.

Composition and Mineralogy of Fra Mauro Breccias

Group F_1 appears to consist largely of soil breccias and is not particularly relevant to the present discussion. However, because the underlying bedrock is the source of the soils much of the following description applies equally well to soil breccias and to the Fra Mauro Formation as represented by groups F_2, F_3 and F_4.

Lithic Fragments

Igneous clasts: At least nine igneous-rock varieties occur in the clastic rocks from Fra Mauro (Wilshire and Jackson, 1972):

    (1)   Intersertal basalt
    (2)   Ophitic basalt
    (3)   Intergranular basalt
    (4)   Variolitic basalt
    (5)   Vitrophyric basalt
    (6)   Graphic quartz-alkali feldspar
    (7)   Plagioclase, orthopyroxene cumulates
    (8)   Hypautomorphic gabbro, norite

Clearly the lunar crust was complex and well differentiated at the time of the Imbrium event.

The most abundant igneous clast type is aphyric intersertal basalt. In thin section these clasts seldom exceed 6 mm in diameter but a single isolated clast of the same lithology that was returned by Apollo 14 is 18 cm in diameter. The intersertal basalts are all rich in plagioclase, which generally has an elongate lath-shaped euhedral form with interspersed subhedral clinopyroxene (Table 5.1). Opaque minerals, largely ilmenite, troilite and metallic iron, occur in small amounts. Very fine-grained apatite (?), ilmenite, troilite, alkali feldspar (?) and glass form the matrix.

TABLE 5.1.

Modal composition of basaltic clasts contained in Fra Mauro clastic rocks (after Wilshire and Jackson, 1972).

	Intersertal					Ophitic				Intergranular	
Plagioclase	60	55	60	58	70	55	40	50	50	80	50
Pyroxene	30	35	--	30	19	39	58	45	45	10	50
Olivine	--	--	--	--	1	5	1	5	5	2	--
Opaque	Tr.	Tr.	1	2	Tr.	1	1	--	--	8	--
Mesostasis	10	10	39	10	10	--	--	--	--	--	--

(Tr. = trace)

Aphyric basalt clasts with an ophitic texture are almost as abundant as the intersertal basalt clasts. The clasts average 5 mm in diameter but some larger individual clasts were returned by the mission. Typically these basalts have lower plagioclase contents than the intersertal basalts but contain a small percentage of olivine. The plagioclase laths are euhedral and penetrate the larger anhedral or subhedral pyroxenes. Olivine, which is sparsely distributed, occurs as ragged grains in cores of pyroxenes. Opaques such as metallic iron, ilmenite and troilite are present in vugs or interstices. Chemically the intersertal and ophitic basalts are distinct (Table 5.2). The ophitic basalts are essentially mare basalts which suggest that mare flooding was already underway at the time the Imbrium Basin was excavated.

Vitrophyric basalt fragments of up to 4 mm in diameter have been recorded in a few rock samples. Some of these fragments are irregular and angular to subangular whereas others are almost perfectly round in form. The rounded fragments (Fig. 5.5) have been described as chondrules by some workers (King *et al.*, 1971. 1972; Kurat *et al.*, 1972; Nelen *et al.*, 1972). Wilshire and Jackson (1972) have identified five subtypes based on the crystal phases: (1) plagioclase only, (2) olivine only, (3) plagioclase and olivine, (4) plagioclase and clinopyroxene and (5) plagioclase, olivine and clinopyroxene. Kurat *et al.* (1972) have made a detailed study of such chondrule-like particles in one breccia (Sample No. 14318). They found considerable variation in the composition of the particles (Table 5.2) with individual particles ranging from anorthositic to basaltic in composition. The vitrophyric basalt fragments quite clearly have a complex source. Kurat *et al.* (1972) have suggested that they are formed by spontaneous crystallization of highly super-cooled, free-floating molten droplets produced during hypervelocity impact. There is reasonable textural information to suggest that this is at least partly true. However, angular particles of similar

Fig. 5.5. A chondrule-like particle in a Fra Mauro breccia. The particle is approximately 450 μm in diameter.

composition and texture are also present in the samples (Wilshire and Jackson, 1972). Furthermore, many of the mineral grains in the clastic rocks are highly rounded and it seems reasonable that some of the rounded vitrophyric basalt fragments may have been produced by mechanical abrasion (Lindsay, 1972a). This last possibility is consistent with the sharp truncation of mineral grains at the surface of some rounded grains, something that is unlikely to occur during spontaneous crystallization in a droplet.

Two other textural varieties of basalt are present as clasts in the breccias — intergranular basalts and variolitic basalts. Neither type is abundant.

Most rocks contain a few clasts of graphic or irregular intergrowths of alkali-feldspar and quartz. Igneous clasts with cumulus textures occur in most of the clastic rocks but are never as abundant as the basaltic rock fragments. Most clasts are less than 2 mm in diameter and typically sub-angular to sub-rounded (Lindsay, 1972a). Wilshire and Jackson (1972) identified seven varieties of cumulate which are extremely varied texturally (Table 5.3). Plagioclase is by far the dominant cumulus mineral.

The hypautomorphic granular textures are generally found in clasts of norite, grabbro or anorthosite. The clasts are up to 7.5 mm in diameter but, because of their coarse grain sizes, representative modes are not obtainable.

TABLE 5.2.

Major element chemistry of basaltic clasts from Fra Mauro Formation breccias.

	Intersertal[a] Basalt (14310)	Aphyric[b] Basalt (14053)	Vitrophyric "chondrules"[c] (14318)				
			1	2	3	4	5
SiO_2	47.2	46.08	45.4	47.2	49.0	45.2	49.5
TiO_2	1.24	2.91	0.05	0.30	0.97	3.4	2.93
Al_2O_3	20.1	12.54	34.1	24.8	17.9	12.1	7.9
FeO	8.38	16.97	0.25	6.0	7.9	17.3	16.6
MnO	0.11	0.255	0.01	0.09	0.15	0.31	0.30
MgO	7.87	8.97	0.34	5.7	9.7	9.3	5.2
CaO	12.3	11.07	18.4	14.9	12.0	11.4	10.6
Na_2O	0.63	0.44	1.19	0.51	0.36	0.34	0.80
K_2O	0.49	0.097	0.11	0.17	0.73	0.41	1.14
P_2O_5	0.34	0.114	0.04	0.11	0.22	0.10	3.0
S	0.02	0.132	---	---	---	---	---
Cr_2O_3	0.18	0.42	<0.02	0.15	0.20	0.37	0.16
Total	98.9	99.998	99.89	99.93	98.98	100.23	98.13

(a) Hubbard *et al.* (1972)
(b) Willis *et al.* (1972)
(c) Kurat *et al.* (1972)

Metamorphic Clasts

There are at least seven types of metamorphic rock which occur as clasts in the Fra Mauro breccias (Wilshire and Jackson, 1972; Chao *et al.*, 1972). Two of the lithologies are metaclastic, two are meta-igneous, a meta-basalt and a metagrabbro, and three are recrystallized textures with a relict clastic texture.

Metaclastic rock fragments are by far the most common clast type (Fig. 5.6). The clasts generally consist of relict angular crystal debris, most of which is plagioclase, but pyroxenes, olivine or quartz also occur. The porphyroclasts form up to 70 per cent of the rock fragment and generally occur in a granoblastic or poikiloblastic matrix. The two metaclastic varieties are separated by color — the dark-colored clasts being more abundant than the light-colored ones. The darker clasts appear to have more opaques than the light ones and the pyroxenes appear to be principally clinopyroxenes.

Opaque minerals are present in the lighter clasts also but occur in larger

TABLE 5.3.

Modal composition (percentages) of cumulate occuring in Fra Mauro breccias (after Wilshire and Jackson, 1972).

Variety	Cumulus Minerals				Post Cumulus Phases				
	Plagioclase	Orthopyroxene	Clinopyroxene	Olivine	Plagioclase	Orthopyroxene	Clinopyroxene	Opaques	Glass
1	45	--	--	--	--	55	--	Tr.	--
2	42	inc. olivine			58	--	--	Tr.	--
3	37	13	--	--	27	27	--	--	3
4	40	--	--	--	--	--	55	5	--
5	79	--	--	--	--	--	15	Tr.	6
6	42	--	--	--	58	--	--	Tr.	--
7	40	inc. olivine (?)			60	inc. opx + alkali feldspar + glass			

Fig. 5.6. A well rounded brown metaclastic-rock fragment in a low grade Fra Mauro breccia. The rounded clast is approximately 900 μm in longest dimension.

clumps. The minerals forming these clasts are difficult to identify; however, Lindsay (1972a) found that the plagioclase/pyroxene ratio of the light clast is lower than for the dark clasts. In thin section clasts are as large as 8 mm. In reality the clasts may have a much greater size range as large clasts in the process of disintegrating are visible in some large rock slabs (Fig. 5.2). Clast relationships are complex and are discussed in detail in a following section.

The light-colored metaclastic rocks are generally coarser grained than the dark clasts and are distinguished by granoblastic or poikiloblastic textures. Plagioclase is the dominant mineral. Orthopyroxene is the dominant mafic mineral although clinopyroxene (probably pigeonite) may dominate in some fragments.

The metabasalt clasts occur in small numbers in most clasts but are never abundant. Prior to metamorphism most of the clasts had an intergranular or intersertal texture. As a result of metamorphism they now consist of a granular aggregate of augite with olivine in some samples, equant opaque minerals, and ragged plagioclases that poikiloblastically enclose other minerals. Metagabbro clasts are rare. A few of the clasts appear to have been troctolites. Both the plagioclase and olivine have been recrystallized to fine-grained granoblastic aggregates but the original gabbroic texture is still evident.

The hornfels clast are essentially monomineralic fragments consisting of plagioclase or olivine, or less commonly of clinopyroxene and ortho-pyroxene. The clasts are generally ovoid and sub-rounded but may be blocky and angular. Generally the size of these particles is less than 3 mm. Plagioclase hornfels are by far the most common and consist of fine interlocking anhedra. A few plagioclase hornfels and most of the other hornfels clasts have granoblastic-polygonal textures. The majority of these aggregates appear to be recrystallized large single crystals, perhaps derived from a coarse-grained gabbro.

Mineral Grains

Plagioclase is the most abundant optically resolvable detrital mineral grain which is present in the Fra Mauro clastic rocks. The grains, which are up to 2.5 mm in diameter, are generally well twinned and in many cases show evidence of shock modification of the twins. Evidence of zoning is present in some grains. Pyroxene grains are pale-brown to colorless augite, colorless pigeonite and orthopyroxenes (Table 5.4). The orthopyroxenes commonly contain clinopyroxene exsolution lamellae. Clinopyroxenes and orthopyroxenes occur in relatively equal proportions although the clino-pyroxenes may be slightly more abundant (Wilshire and Jackson, 1972; Lindsay, 1972a). Olivine is not common as single mineral grains in any

TABLE 5.4.

Major element chemistry of detrital mineral phases in an F_4 Fra Mauro breccia (from Brunfelt et al., 1972).

| | OLIVINES | | | | | | |
	1	8	14	16	17	18	20
SiO_2	37.78	38.72	39.48	40.65	37.84	40.34	37.59
TiO_2	0.0	0.0	0.0	0.0	0.0	0.0	0.0
Cr_2O_3	0.13	0.12	0.15	0.16	0.13	0.07	0.11
Al_2O_3	0.38	0.54	0.21	0.39	0.41	0.39	0.46
FeO	30.21	23.44	18.36	13.55	28.03	15.17	29.30
MnO	0.32	0.23	0.18	0.13	0.30	0.14	0.28
MgO	28.97	37.00	41.67	45.80	32.96	44.01	32.38
CaO	0.51	0.36	0.0	0.06	0.39	0.15	0.25
Na_2O	0.0	0.0	0.0	0.0	0.0	0.0	0.0
	98.30	100.41	100.05	100.74	100.06	100.27	100.37

| | PIGEONITES | | | | | SPINEL 14303.49 |
	3	4	12	13	19	
SiO_2	52.72	53.15	54.04	48.58	50.29	0.0
TiO_2	0.43	0.38	0.31	0.39	0.31	0.0
Ca_2O_3	0.80	0.82	0.84	0.80	0.30	5.61
Al_2O_3	2.00	1.91	1.57	1.17	0.82	62.91
FeO	16.56	15.55	16.78	17.85	30.72	13.41
MnO	0.34	0.33	0.34	0.27	0.55	0.0
MgO	20.57	21.60	23.34	24.74	10.75	18.17
CaO	5.11	5.15	3.39	5.15	5.98	0.0
Na_2O	0.0	0.0	0.0	0.0	0.0	—
	98.53	98.89	100.61	98.95	99.76	100.10

| | ORTHOPYROXENES | | | | CLINOPYROXENES | | |
	7	9	10	11	2	6	15
SiO_2	54.30	55.81	56.77	54.50	52.31	51.00	52.73
TiO_2	0.33	0.35	0.17	0.65	0.58	1.50	0.91
Cr_2O_3	0.34	0.56	0.29	0.33	0.01	0.65	0.18
Al_2O_3	1.18	0.98	0.90	1.16	1.56	1.79	1.42
FeO	17.91	12.02	11.30	18.63	9.77	8.47	15.64
MnO	0.30	0.21	0.22	0.30	0.22	0.20	0.30
MgO	23.58	29.31	30.38	22.16	14.50	16.10	12.65
CaO	1.79	1.74	0.68	2.25	21.12	19.80	16.34
Na_2O	0.0	0.0	0.0	0.0	0.0	0.0	0.11
	99.73	100.98	100.71	99.98	100.15	99.56	100.28

sample (Fig. 5.7) and the few grains are generally small and angular although some grains reach 2.3 mm in diameter. The composition of the olivines is variable but in general they tend to be magnesian (Table 5.4). Minor mineral constituents include ilmenite, metallic iron, troilite, chromian spinel, ulvöspinel, native copper, armalcolite, zircon, apatite and a mineral tenatively identified as K-feldspar and dark-green amphibole (Lunar Sample Preliminary Examination Team, 1971; Wilshire and Jackson, 1972; Chao *et al.*, 1972; Lindsay, 1972a).

The Matrix and Metamorphism

The matrix materials (materials <25 μm) of the Fra Mauro breccias show, in varying degrees, considerable evidence of chemical re-equilibration. On the basis of the abundance of glass and the matrix texture the breccias may by divided into 8 groups which fit into three metamorphic grades (Warner, 1972) (Table 5.5). The groups are numbered 1 through 8 with higher numbers indicating higher metamorphic grade. The lower grades are distinguished by (1) decreasing matrix glass and (2) decreasing numbers of glass clast. Although Warner (1972) includes grades 1 and 2 with the Fra Mauro Formation metamorphic series, the textural evidence suggests that they are much younger rocks formed by the impact cratering of lunar soil on the surface of the Fra Mauro Formation (Chao *et al.*, 1972; Engelhardt *et al.*, 1972; Lindsay, 1972a). A further increase in metamorphic grade produces increasingly more recrystallized equant or granoblastic matrix textures, until

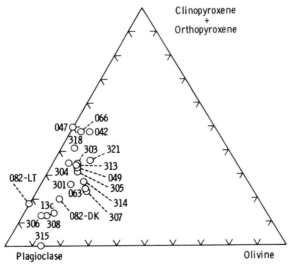

Fig. 5.7. Triangular plot of plagioclase-pyroxene-olivine microclasts in Fra Mauro breccias (after Wilshire and Jackson, 1972).

TABLE 5.5.

Metamorphic grades based on Warner (1972) and Williams (1972).

°C	Grade	Warner Group	Glass Content	Matrix Texture	Plagioclase Composition	Pyroxene Composition	Process
600	Soil	Soil					
700		0					
800	Low	1	High	Detrital			Devitrification
900		2	Intermediate				
	Medium	3	Low	Equant		En_{61}	
1000		5	None			En_{55}	Pyroxene becomes equant
		6 + 4			$An_{83}\ Or_1$	En_{63}	
1100	High	7		Euhedral	$An_{76}\ Or_2$	En_{68}	Plagioclase becomes equant
		8		Sheath-like	$An_{73}\ Or_7$	En_{72}	Matrix Melts

finally at the highest grade the texture becomes sheath-like (Fig. 5.8).

Along with the metamorphic grades the chemistry of the matrix materials shows trends consistent with the re-equilibration of the detrital components. The composition of plagioclases in the lower grade breccias displays a continuum over a wide compostional range (An_{64} to An_{99}) (Fig. 5.9). With increasing metamorphic grade the spread of the compositional range not only decreases but the mean plagioclase composition becomes less anorthositic as it equilibrates. In the higher grades (6, 7, and 8) the orthoclase content increases systematically due to the dissolution of small K-feldspar blebs from the matrix into the plagioclase. Larger grains of plagioclase never fully equilibrate but develop rims, all with essentially the same composition regardless of the composition of the original grain. This suggests that the rims are authigenic overgrowths rather than reaction rims (Warner, 1972; Lindsay, 1972a). The implication is that the Fra Mauro breccias were not subjected to high temperatures for a long time period.

The pyroxene mineralogy produces a similar picture. Lower grade breccias show a wider compositional range in keeping with their detrital origins. The higher grade breccias display preferred orthopyroxene compositions. Compositions change from En_{55} for Group 5 to En_{72} in Group 8.

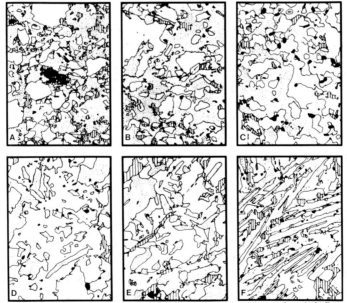

Fig. 5.8. Texture of Fra Mauro breccias of different metamorphic grades. (A) Group 1, (B) Group 4, (C) Group 5, (D) Group 6, (E) Group 7, and (F) Group 8. Longest dimension of sketch 190 μm. Clear areas are pyroxene + olivine, dotted areas are feldspar + glass, black areas are opaques and lined areas are vugs and polishing artifacts (from Warner, 1972, *Proc. 3rd Lunar Sci Conf., Suppl. 3, Vol. 1,* MIT/Pergamon).

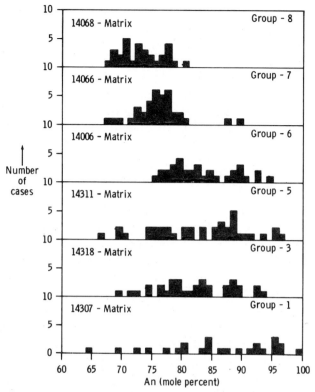

Fig. 5.9. Variations in the An content of matrix plagioclase in Fra Mauro breccias of different metamorphic grade (from Warner, 1972, *Proc. 3rd Lunar Sci. Conf., Suppl. 3, Vol. 1*, MIT/Pergamon).

Pyroxenes in higher grade breccias, as with the feldspars, show the development of rims with a constant composition. The pyroxene rims are difficult to evaluate because zoning is common in the mineral. However, pyroxenes normally zone outward to Fe-rich subcalic augite whereas the rims in this case consist of Mg-rich orthopyroxene suggesting that like the feldspars, the pyroxene rims are authigenic overgrowths. Matrix olivine, like plagioclase and pyroxene, shows a compositional spread in the lower grade breccias with a trend towards more homogeneous compositions in higher grades (Fig. 5.10). As with the pyroxenes the olivine equilibrium composition becomes more Mg-rich. In the higher metamorphic groups the larger detrital olivine grains have mantles (Cameron and Fisher, 1975) which characteristically have two zones: an inner corona of pyroxene, ilmenite and commonly plagioclase and an outer light colored halo where the matrix is depleted in ilmenite (Fig. 5.11). Unlike the pyroxene and plagioclase rims the olivine mantles are reaction rims involving a matrix to corona diffusion

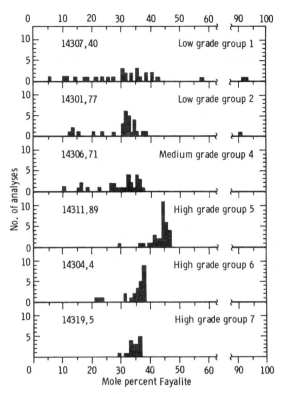

Fig. 5.10. Composition of detrital olivines in Fra Mauro breccias of different metamorphic grades (after Cameron and Fisher, 1975).

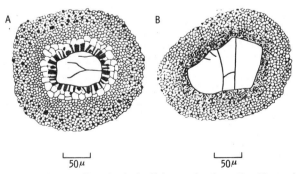

Fig. 5.11. Reaction coronas surrounding detrital olivine grains in a Fra Mauro breccia. Ilmenite is black; all silicates are white. (A) The core of olivine is surrounded by a corona containing an inner zone of pigeonite + ilmenite and an outer zone of coarse pigeonite devoid of ilmenite, a fine-grained halo devoid of ilmenite and finally normal matrix material with ilmenite. (B) The core of olivine is surrounded by prismatic orthopyroxene disseminated ilmenite and intersertal plagioclase and finally normal matrix material with a granoblastic texture (from Cameron and Fisher, 1975).

of TiO_2, and a corona to matrix diffusion of MgO and FeO.

The higher metamorphic grades of Fra Mauro breccia contain vugs lined with crystals (McKay *et al.*, 1972). The vugs may be either irregular in shape and interconnected in a network, or spherical in shape and apparently without interconnections. The latter type could be more correctly described as vesicles. Interconnected pores within the breccias account for about 4 per cent of the volume of the rock. This permiability would readily allow hot vapors to pass through the rock and it has been suggested that the minerals lining the vugs are the product of a metamorphic hot-vapor phase. The vapor phase appears to have contained oxides, halides, sulfides, alkali metals, iron and possibly other chemical species. Passage of this vapor through the breccias resulted in the deposition of crystals of plagioclase, pyroxene (Fig. 5.12), apatite (Fig. 5.13), ilimenite, iron (Fig. 5.14), nickle-iron, and troilite that extend from the vug walls and bridge open spaces. Similar vapor phase products have been encountered in high grade breccias from the Apollo 15 landing site at the foot of the Apennine Mountains (Fig. 5.15) (Clanton *et al.*, 1973).

The Metamorphic Environment

Williams (1972) assembled a large body of evidence to establish a metamorphic temperature scale for the Fra Mauro breccias. The temperature scale agrees well with the metamorphic grades established by Warner (1972), except that it was found necessary to eliminate group 4 and combine it with group 6 while at the same time creating a group 0 for another type of as yet unreported breccia (Table 5.5).

The proposed model involves the deposition of the Fra Mauro Formation in two layers. The basal layer was deposited with a temperature of at least 1100°C and possibly as high as 1300°C. The upper layer had an initial temperature of less than 500°C. Williams (1972) has suggested that these two layers may have been deposited simultaneously in a situation somewhat similar to laminar and turbulent zones in terrestrial density currents. However, this seems very unlikely as a base surge is almost certainly very turbulent throughout and except for sorting effects it most likely will be very-well homogenized. It seems more likely that the two layer effect is exaggerated by the inclusion of more recently formed low temperature soil breccias in the Fra Mauro cooling sequence. Many of the low grade breccias included in the Fra Mauro Formation sequence contain abundant textural information (for example, agglutinates) which reflect a long exposure history in the lunar soil (Lindsay, 1972a).

The Williams model suggests that the cooling of the lower zone from

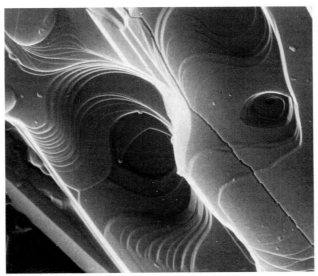

Fig. 5.12. Growth steps on the surface of a pyroxene crystal deposited by a vapor phase during thermal metamorphism. The longest dimension of the photo is 8 μm (NASA Photo S-73-30449).

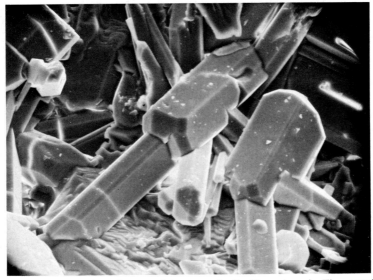

Fig. 5.13. An aggregate of apatite crystals in a vug of a Fra Mauro breccia. The crystals are believed to be vapor phase deposits resulting from thermal metamorphism (McKay *et al.*, 1972) (NASA Photo S-72-17292). The field of view is approximately 70 μm wide.

Fig. 5.14. A perfect metallic iron crystal deposited on a pyroxene substrate during thermal metamorphism by a vapor phase (Clanton *et al.*, 1973). The edges of the cube are 2 μm in length (NASA Photo S-72-53029).

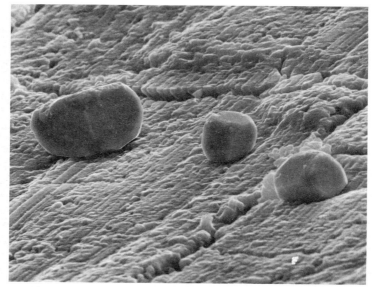

Fig. 5.15. Three small metallic iron crystals on a pyroxene substrate. The iron crystals were deposited by vapor phase crystallization (Clanton *et al.*, 1973). The edges of the largest crystal are 6 μm long (NASA Photo S-72-53357).

1200°C to 900°C occured in minutes while the middle layers may have taken days to cool to 800°C. The model produces a refining effect which drives volatiles such as CO, CO_2, H_2, and rare gases from the lower part of the flow (which is being rapidly sealed by sintering) into the still porous upper layers. The volatiles may either be driven from the flow entirely or simply be absorbed at higher levels in the flow. The metal content in the various grades may also relate to vapor phase transfer (Fig. 5.15) (McKay et al., 1972; Clanton et al., 1973). However, caution should be exercised in interpreting the high metallic iron contents in low grade breccias (Gose et al., 1972) in the same manner because metallic iron is produced by micrometeoroid reworking of the lunar soils (Gose et al., 1975). That is, soil breccias may be expected to contain relatively larger amounts of metallic iron. An alternative to vapor phase transfer of metals is in situ reduction by the vapor phase (Williams, 1972). Data are not available to determine the temperature history below 800°C. It is certain, however, that group 8 rocks were heated to temperatures above 1100°C and that group 1 rocks require no more than 800°C. Finally most of the volatiles could be expected to have been redistributed within 24 hours. This combined with the earlier discussion of the mechanics of base surges (Chapter 3) suggests that the Fra Mauro Formation may have been excavated and transported, deposited and metamorphosed in less than two days — a very short time by terrestrial sedimentation standards.

Sedimentary Textures

There is an abundance of textural information in the Fra Mauro breccias that provides considerable insight into the environment of transportation and deposition of the rocks. In general these observations support Gilbert's (1893) original hypothesis that the Fra Mauro Formation is an ejecta blanket excavated from the Imbrium Basin.

Interclast Relationships

One of the most conspicuous features of the Fra Mauro breccias is an elaborate series of clast-within-clast relationships which can be viewed as a kind of stratigraphic history of erosional and depositional episodes. In Table 5.6 these complex interrelations are shown schematically. These data have been taken from Wilshire and Jackson (1972) and rearranged in terms of Warner's (1972) metamorphic groups. Higher metamorphic grades should come from deeper within the Fra Mauro Formation. The notable feature of this arrangement is the increased complexity of clasts in higher grades (greater depths). In conjunction with this there is a general increase in the

TABLE 5.6.

Interclast relationships for Fra Mauro breccias. Data from Wilshire and Jackson (1972) rearranged in terms of Warner's (1972) metamorphic grades with soil breccias (Group 1) excluded. Note that clast complexity increases in higher metamorphic grades indicating cycles of erosion and re-deposition at the base of the base surge which deposited the formation (Lindsay, 1972a).

Grade	Sample No.	Main Clast Type	Host Clast	Clasts in 1	Clasts in 2	Clasts in 3	Percentage of clasts >1 mm
			1	2	3	4	
2	14301	C^*_1	C^*_1	I_2, I_5, I^*_7 $I^*_8 C_1^*$			18
3	14063	C^*_2	C^*_2	C^*_1, I^*_7			23
	14315	C^*_1, C^*_2		G_2, C^*_2, I_3			20
6	14303	C^*_2	C^*_2	C^*_1, I_5			28
	14304	C^*_2	C^*_2	C^*_1, I^*_7, I^*_8			25
	14305	C^*_2	C^*_2	C^*_1, I^*_7			28
	14306	C^*_2	C^*_2	C^*_2, I_1, I^*_7	C^*_1, C^*_2		15
7	14308	C^*_2	C^*_1, C^*_2	C^*_1	C^*_2	I^*_8	20
	14312	C^*_2	C^*_2	C^*_2			45
	14314	C^*_2	C^*_2	$C^*_1, C^*_2,$ I_6, I^*_8			25

* Metamorphosed lithology

Basalts

I_1 Intersertal
I_2 Intergranular
I_3 Ophitic
I^*_4 Metabasaltic

Clastic Fragments

C^*_1 Light-colored metaclastic rock
C^*_2 Dark-colored metaclastic rock

Coarse Grained Igneous

I_5 Graphic quartz-alkalic feldspar
I_6 Cumulates
I^*_7 Recrystallized plagioclase
I^*_8 Recrystallized olivine

Glass

G_1 Pale-yellow to brown
G_2 Colorless

proportion of clasts with increasing metamorphic grade (or depth).

Clastic lithic fragments can not always be readily separated visually from the containing matrix (Lindsay, 1972a). Dark clastic-lithic fragments, where they have well-defined boundaries, are found to be very angular. However, the boundaries of these clasts are frequently diffuse and their shapes are distorted plastically. Dark-grey clastic fragments occur in all stages of disintegration, from clasts with slightly diffuse boundaries to those which occur as sinuous wisps of darker material in a light matrix. Some dark lithic clasts appear to be frozen midway in the process of disintegrating. Chips and slivers of the clast occur floating in the light-colored matrix a short distance from the point of origin on the larger clast (Lindsay, 1972a). Similar textures are seen on a larger scale in blocks of breccia returned by Apollo 14 (Fig. 5.16). Overall the textures are very similar to those encountered in terrestrial turbidites and mass-movement deposits, as a result of penecontemporaneous erosion of a semi-consolidated plastic substrate (Crowell, 1957; Dott, 1961; Lindsay, 1966; Lindsay et al., 1970). These relationships, in combination with the fact that dark-clast breccias appear to be more abundant deeper in the Fra Mauro Formation, suggest that a proportion of the interclast relationships result from alternating phases of erosion and deposition by the base surge which deposited the formation. The early phases of the flow appear to have had a darker matrix than the later phase. Such behavior is expected in a fluid travelling in supercritical flow, particularly if the terrain is irregular and if antidunal structures are formed (see Chapter 3).

Not all of the dark clastic-rock fragments can be explained in this way. Some fragments appear to have undergone an earlier metamorphic episode. These clasts are also relatively-well rounded, suggesting a history of post-metamorphic abrasion during transport. They are generally contained

Fig. 5.16. A grade 6 clastic rock (14321) from the Fra Mauro Formation. Dark clastic-rock fragments appear to be frozen in the process of disintegrating and suggest penecontemporaneous erosion in a large-scale base surge. The large dark clast is 2 cm in longest dimension (NASA Photo AS-72-17892).

within the dark-colored clastic rock fragments (Lindsay, 1972a). Clearly they belong to one or more earlier generations of impact generated and metamorphosed clastic rocks. One likely candidate as a source for these earlier metaclastic fragments is not in the Serenitatis Basin ejecta blanket, which must have provided some materials for the later Imbrium event. However, the outer crust of the moon must be exceedingly complex and it is possible that the clasts could have come from any number of earlier events.

The Fra Mauro Formation clasts are evidently the product of a complex history of multiple brecciation and metamorphism (Wilshire and Jackson, 1972) with a strong overprint of sedimentary textures resulting from penecontemporaneous erosion and deposition during transport (Lindsay, 1972a).

Grain Size of the Fra Mauro Breccias

The clastic rocks of the Fra Mauro Formation are in general fine-grained and very-poorly sorted (Lindsay, 1972a). The median grain size ranges from 4.32ϕ (50 μm) to 8.65ϕ (2.5 μm). The standard deviation of the size distributions is large; generally of the order of 4.5ϕ. The grain size of the rock samples decreases with increasing distance from Cone Crater, which suggests that samples from deeper in the Fra Mauro Formation are coarser grained (Fig. 5.17). Weak size grading is consistent with a base surge origin for the Fra Mauro Formation. It was also found that the dark clastic rock

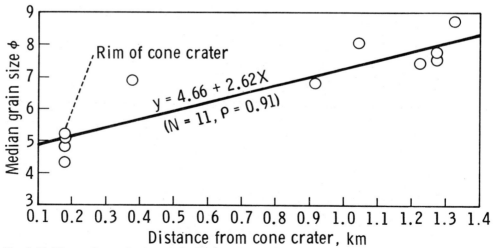

Fig. 5.17. The median grain size of clastic rocks from the Fra Mauro Formation decreases away from the rim of Cone Crater. By analogy with terrestrial impact craters this suggests that the Fra Mauro Formation is a graded unit. That is, detrital materials are coarser deeper in the formation (after Lindsay, 1972a).

fragments which are presumed to be penecontemporaneously eroded are finer grained than matrix materials in which they are embedded. It thus seems possible that the earlier phases of the base surge which deposited the Fra Mauro Formation consisted of finer-grained detrital materials.

Roundness of Detrital Materials

When one thinks of increased roundness in terrestrial terms it is almost automatic to associate the variable with the action of water. Consequently, the fact that many detrital particles in the Fra Mauro rocks show relatively high degrees of roundness was unexpected (Lindsay, 1972a). The older metaclastic-rock fragments have a mean roundness of 0.51 (Waddell scale); other lithic clasts have lower mean roundness values. There is considerable variation among the lithologies suggesting that some lithologies either travelled further than others or were subjected to a different abrasion history. For example, many of the feldspathic clasts are relatively angular whereas the olivine and pyroxene-rich clasts tend to be better rounded.

In terms of sampling, the most complete roundness data were obtained from the smaller but more abundant grains of plagioclase and pyroxene (Fig. 5.18). The mean roundness of plagioclase grains from various samples ranges

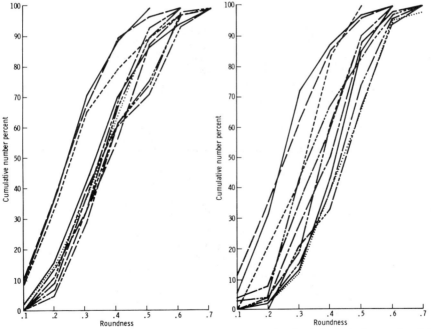

Fig. 5.18. Cummulative roundness frequency distribution for detrital plagioclase grains (left) and pyroxene grains (right) in clastic rocks from the Fra Mauro Formation (after Lindsay, 1972a).

from 0.250 to 0.379. Pyroxene grains tend to to better rounded than the plagioclase and have mean values ranging from 0.261 to 0.435. Individual mineral grains of either species may however range from 0.1 to 0.6. Despite the higher roundness of the pyroxene grains the roundness values of the two minerals have a strong sympathetic relationship. This suggests that both have undergone the same abrasion history but that the pyroxenes have a higher inherited roundness. The roundness sorting for the two minerals is relatively small; 0.111 to 0.145 for the plagioclases and 0.095 to 0.131 for the pyroxenes. This suggests prolonged application of abrasive energy under a relatively uniform set of conditions. However, there are extreme roundness values in both mineral species and well rounded but broken grains of plagioclase are present in most samples (Fig. 5.19) suggesting that the depositional environment was highly energetic. It thus seems reasonable to suggest that much of the kinetic energy of the base surge was converted to heat energy as a result of internal particle interaction (Anderson *et al.*, 1972).

Maturity of Fra Mauro Lithologies

In its most general sense the term "maturity", when applied to terrestrial sedimentary rocks, implies some measure of energy released in producing the final detrital assemblage. According to Pettijohn (1957) "the

Fig. 5.19. A broken plagioclase grain which was previously rounded to a high degree. Such relationships suggest a highly energetic and abrasive transportational mechanism with intense particle interaction. The longest dimension of the particle is approximately 500 μm (after Lindsay, 1972a; NASA Photo S-71-30839).

mineralogic maturity of a clastic sediment is the extent to which it approaches the ultimate end product to which it is driven by the formative processes that operate upon it." In the terrestrial environment it is generally convenient to measure maturity by looking at the relative abundance of some inert mineral such as quartz in relation to less stable feldspars and mafic minerals. However, such concepts are dependent upon the fact that water dominates the terrestrial environment in terms of both the physical and chemical properties of the detrital materials. On the moon quartz is unimportant as a detrital mineral and in the absence of water feldspars and pyroxenes are extremely stable. Consequently, a somewhat different approach is called for (Lindsay, 1972b).

The depositional environment of the Fra Mauro Formation was extremely energetic and considerable particle interaction must have occured. This resulted in the disaggregation of larger lithic clasts and intense abrasion of detrital particles such that, with increasing distance of travel, Fra Mauro breccias should be less lithic, more mineralic and contain more matrix material. This suggests that the maturity of major ejecta blanket lithologies can best be determined by looking at the relative proportion of lithic and mineral clasts to matrix material. In Figure 5.20 the Fra Mauro lithologies are plotted on a ternary diagram using these end members. Since there is no water on the moon to selectively remove the fine fractions, as is generally the case on earth, a mature breccia contains more matrix material (<15 μm)

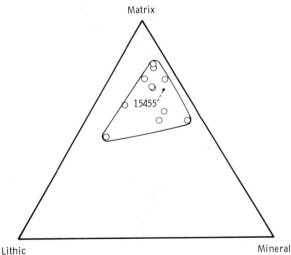

Fig. 5.20. Ternary plot of Fra Mauro rocks to indicate relative maturity of the sample. Mature rocks contain an abundant matrix. Sample 15455 comes from the Apennine Front at the Apollo 15 site, and may have been excavated by the Imbrian event from the depositional unit surrounding the Serenitatis Basin (after Lindsay, 1972b).

than an immature breccia. There is some scatter in the samples but most cluster towards the matrix apex indicating a considerable distance of transport which is in keeping with the known situation. A crystalline breccia from the Apennine Front at the Apollo 15 site is plotted for comparison. It contains well-rounded mineral clasts and is obviously mature which is not consistent with its position relative to the Imbrium Basin. This suggests that it comes from an older ejecta blanket perhaps from the Serenitatis event.

Genesis of the Fra Mauro Formation

The Fra Mauro Formation appears to be the product of a large scale base surge produced during the excavation of the Imbrium Basin. The flow moved radially from the basin at velocities of the order to 23 to 32 m s^{-1}. For most of its travel distance the flowing mass moved in supercritical flow producing a hummocky terrain which may relate to antidunal structures. The debris flow was extremely turbulent and highly energetic, resulting in the abrasion and rounding of the detrital particles and the general size grading of the resulting Fra Mauro Formation sediments. The earlier phases of the flow appear to have consisted of finer-grained materials with a darker matrix than later phases. The flow appears to have gone through alternate cycles of deposition and re-erosion, resulting in complex clast interrelationships much like mud-chip breccias in terrestrial mass-movement deposits. The suggests that the flow contained a high volume concentration of solid materials and that it had a high dynamic viscosity. The cycles of deposition and re-erosion probably relate to the balance between inertial and gravitational forces which result in the growth and development of antidune structures with a characteristic wavelength.

The flow had a high temperature; at least 1100 to 1300°C at the time of deposition. These high temperatures resulted in large part from the direct release of heat energy during hypervelocity impact. However, significant amounts of heat must have been generated later by particle interaction during transport. The high temperatures at the time of deposition of the Fra Mauro Formation led to very-rapid thermal metamorphism of the breccias, largely in the form of matrix recrystallization and vapor phase transferral.

The lithic clasts contained in the Fra Mauro Formation are of great variety, suggesting a very-complex source area. The igneous clasts, which include basalts of several varieties distinct from the mare basalts, suggest igneous crystallization and differentiation at depth. Metaclastic rock fragments, the dominant clast type, suggest earlier cycles of large scale impact fragmentation and thermal metamorphism. One source of these earlier metaclastic fragments may have been the Serenitatis Basin ejecta blanket, although the complexity of clast interrelations is such that even earlier events must also be represented. The pre-Imbrium terrain, presumed to have existed 3.95 aeons ago, was already extremely complex. Multiple ejecta blankets similar to the Fra Mauro Formation already existed, a complex suite of basalts had been erupted and incorporated into the ejecta blankets and igneous differentiation and intrusion had occured (Wilshire and Jackson, 1972; Lindsay, 1972a). The processes operative in the pre-Imbrium time were thus not unlike those operative later, suggesting that a primordial anorthositic crust is unlikely. Rather the Fra Mauro Formation represents a long history of impact and ejection that records the first half aeon of planetary accretion, of which the Imbrium event was one of the last (Wilshire and Jackson, 1972).

Lithology of the Cayley Formation

The Cayley Formation has long attracted attention because of its distinctive appearance and distribution among the surrounding dark mare. The Cayley Formation and other similar smooth plains units of Imbrian age are widespread and occur in local depressions of most lunar highlands. Consequently an understanding of the nature of the Cayley Formation is important in understanding the overall geology of the moon. The main objective of the Apollo 16 mission was to investigate the Cayley Formation at the western edge of the Descartes Mountains about 50 km west of the Kant Plateau.

Initially the Cayley Formation was included in the Apenninian Series with the hummocky deposits of the Imbrium event (Eggleton and Marshall, 1962). The hummocky deposits of the Fra Mauro Formation become continually smoother southwards and appear to grade into plains units. However, the plains units locally have sharp contacts with the hummocky Fra Mauro Formation, which caused Wilhelms (1965) to isolate the Cayley Formation from the Fra Mauro Formation and to suggest that the Cayley Formation was volcanic in origin. The volcanic hypothesis remained in good standing until the Apollo 16 landing (Milton, 1972; Trask and McCauley, 1972; Wilhelms and McCauley, 1971; Elston et al., 1972), but, the samples returned by this mission failed to provide any information to uphold this hypothesis. Almost all of the returned samples were impact-generated breccias with an equally complex history as that of the Fra Mauro breccias.

Classification of the Cayley Formation Lithologies

The Apollo 16 crew were able to identify two main breccia types at the

Descartes landing site: (1) fragmental rocks with dark clasts in a friable white matrix and (2) fragmental rocks with white clasts in a very coherent dark matrix. These observations were used to form the basis of a megascopic classification in which five main breccia groups are recognized on the basis of the proportion of light- and dark-grey clasts and matrix color (Wilshire *et al.*, 1973). Many of the rock samples are partly enveloped in or completely veined by glass which was presumably formed during the excavation of two nearby craters of medium size called North and South Ray craters for the purposes of the mission. The vitrification and the mild brecciation accompanying it have not been considered in the classification. The majority of the rocks are polymict, but, unlike the Fra Mauro lithologies, some rocks are monomict. The clasts are equant to highly irregular in shape and as in the Fra Mauro lithologies, range from angular to well rounded, but equant subangular clasts predominate.

Although there is a considerable variety of clast types two are clearly dominant (1) dark-gray aphanitic to finely-crystalline metaclastic rocks and (2) white to light-gray, partly-crushed to powdered, feldspathic rocks. Other clast types include light-gray or white lithologies with a granoblastic texture, a variety of gabbroic to anorthositic rocks with medium to coarse grain size, and rare feldspar-poor basaltic clasts. The matrices of the light- and medium-gray matrix breccias are generally relatively friable and not obviously altered by thermal events. In contrast, the dark matrix breccias are coherent as a result of thermal metamorphism and at times fusion. The relationships among the five groups suggested by Wilshire *et al.* (1973) are shown in Figure 5.21.

Type 1 rocks contain light-gray lithic clasts in excess of dark-gray lithic clasts and have light-colored matrices. This lithology forms 20 per cent of the clastic-rock samples returned by Apollo 16. The abundance of clasts (particles >1 mm) in these rocks is low, and dark clasts are almost absent. Mineral clasts are the most common and consist largely of plagioclase. Type 1 rocks are apparently monomict breccias, the clasts of which are simply uncrushed rock material. The rocks are generally relatively friable and appear unmetamorphosed.

Type 2 rocks contain dark lithic clasts in equal or greater numbers than light lithic clasts and they have light matrices. This rock type forms 38 per cent of the returned Apollo 16 clastic rocks and is the most abundant clastic lithology. Clasts are generally more abundant in these rocks than in type 1 rocks and many rocks contain both dark and light clasts. Mineral clasts are common, and as in type 1 rocks plagioclase dominates. Most of the rocks are polymict but a few are monomict. Unlike the type 1 lithology the matrix of the type 2 rocks does not have the same provenance as the lithic clasts. Type

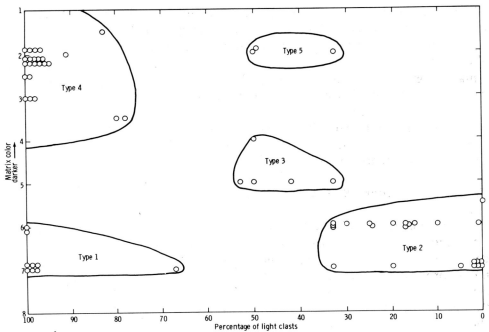

Fig. 5.21. Classification of Cayley Formation breccias returned by Apollo 16 (after Wilshire, H. G., Stuart-Alexander, D. E. and Jackson, E. D., *Jour. Geophys. Res.*, *78*: 2379-2392, 1973, copyrighted American Geophysical Union).

2 lithologies are friable and apparently unmetamorphosed.

Type 3 lithologies are intermediate in character with relatively equal proportions of light and dark clasts and medium-gray matrices (Fig. 5.21). Next to the type 5 lithologies these are the least abundant rock type and form only 7 per cent of the samples returned by Apollo 16. The abundance of clasts is moderate, and all contain some light and dark lithic clasts. Mineral clasts are not as abundant as in types 1 and 2. All of these rocks are polymict and contain a noticeably wider range of lithologies than all other lithologic types. Clast types include glass, brown- and green-pyroxene gabbros and spinel-bearing plagioclase rocks as well as the lithic clasts. The coherence of the type 3 breccias is variable. The texture of these rocks suggests that they are soil breccias formed at a later time on the surface of the Cayley Formation. Soil breccias are treated in more detail in Chapter 6.

Light-gray lithic clasts predominate in type 4 breccias which also have dark matrices. These rocks are second only to type 2 lithologies in abundance and form 31 per cent of the Apollo 16 clastic rock sample. The abundance of clasts in these rocks is extremely variable although most contain only light clasts with a few or no dark clasts. Mineral clasts are considerably less common than in types 1, 2 and 3 lithologies. The rocks are extremely-fine grained and have coherent finely-annealed matrices.

Type 5 rocks form only 4 per cent of the clastic rocks and are the least abundant lithology returned by Apollo 16. They contain dark-lithic clasts in equal or greater proportions than the light-lithic clasts and have dark matrices. Clast abundance is high and all contain both light- and dark-lithic clasts. The dark clasts are very similar to the rock matrix. Mineral clasts are present but not abundant. The rocks are all polymict but the variety of the clast lithologies is not great. The rocks are highly coherent and have finely-annealed matrices and some well developed gas cavities (Fig. 5.22).

Relationships Among Breccia Types

The breccias collected from the surficial debris of the Cayley Formation appear to consist of two principal components: (1) coherent dark-gray fine-grained materials which generally show evidence of thermal metamorphism and at times fusion and (2) friable light-gray fine-grained clastic materials that generally show no evidence of thermal metamorphism (Wilshire *et al.*, 1973). These basic materials are the same whether they form clasts or the matrix of a particular breccia.

The dark clasts of type 2 lithology are indistinguishable from the matrices of lithology types 4 and 5. Likewise the white clasts of type 4 breccias are similar to type 1 breccias and the matrices of type 2 breccias. This suggests that the main breccia types except for type 3 breccias are all genetically related in some complex manner. Type 3 breccias are soil breccias and are ultimately derived by the reworking of all other bedrock lithologies.

Textural evidence from transitional lithologies suggests that type 4 breccias provide the source materials from which types 1, 2 and 5 are derived (Wilshire *et al.*, 1973) (Fig. 5.23). Type 2 breccias are derived directly from the type 4 breccias by a simple rebrecciation during later impact events. The dark-gray aphanitic matrix of the type 4 breccias is tough and resistant to crushing so that it breaks into large angular clasts. The unannealed white clasts on the other hand are readily disaggregated to form the matrix of the new lithology. Type 5 breccias may be produced by rebrecciating clast-poor parts of the original type 4 lithology. Finally the type 1 breccias, or cataclastic anorthosites as they have been called, may be formed from type 4 clasts.

The upper layers of the Cayley Formation have undergone extensive reworking and it is possible that none of the original first cycle breccias have survived intact. The textural complexity of the type 4 breccias suggests that they may also be the end product of multiple brecciation events. It seems

Fig. 5.22. Some representative clastic rocks from the Apollo 16 site. (A) and (B) Type 2 breccias, (C) a
Type 5 breccia (NASA Photos S-72-37214, S-72-37758, S-72-37155).

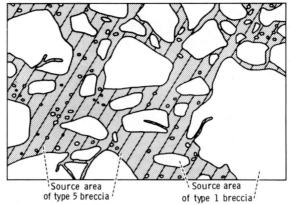

Fig. 5.23. Drawing showing a hypothetical terrain of type 4 -like breccia with white clasts in a dark fine-grained matrix. Rebrecciation of such rock could provide (1) unaltered type 4 breccia, (2) type 2 breccia as a consequence of the friable character of unannealed clasts such that the clasts are reduced to matrix grained size before the annealed dark matrix, (3) type 1 breccia as a consequence of isolating pieces of large clasts from the matrix, and (4) type 5 breccia as a consequence of isolating shattered clast-poor parts of the matrix (from Wilshire, H. G., Stuart-Alexander, D. E. and Jackson, E. D., *Jour. Geophys. Res.*, 78: 2379-2392, 1973, copyrighted American Geophysical Union).

reasonably clear that, with the possible exception of the soil breccias, all of the clastic rocks from the Apollo 16 site overlying the Cayley Formation have a common provenance. The mineralolology in the following section has therefore been generalized.

Petrology of Cayley Formation Lithologies

The Dark Matrix

The matrix of the type 4 breccias, from which dark clasts and matrix materials of other rocks appear to be derived, consists predominately of high-Ca plagioclase granules (An_{91}) (20-25 μm long) and low-Ca pyroxene (En_{71}) and olivine granules (Fo_{75}) (10-20 μm long) (Table 5.7). Most of the low-Ca pyroxene is oriented into oikocrysts of 100 μm diameter (Albee *et al.*, 1973). Ilmenite plates form sheaf-like aggregates.

Vugs form about 3 per cent of the matrix; they are surrounded by a matrix rich in plagioclase and poor in low-Ca pyroxene and contain projecting networks of crystals or pyroxene, ilmenite, plagioclase and apatite similar to the Fra Mauro Formation lithologies. K-rich material occurs as small patches (<15 μm) throughout the matrix and appears to be detrital glass.

Opaque minerals include armalcolite, ilmenite, Cr-rich spinel, rutile, baddeleyite, metallic iron, troilite and schreibersite. The armalcolite occurs either as small independent grains or in a complex association with Cr-rich spinel, rutile and baddeleyite. Two compositional varities of armalcolite appear to be present. The mineralogy and textural relationships are on the whole similar to high-grade Fra Mauro breccias.

Mineral Grains

Like the Fra Mauro lithologies the larger mineral grains in the Cayley rocks are not in equilibrium with the enclosing matrix. The plagioclase compositions in type 4 breccia matrices fall into two groups: (1) the fine-grained matrix plagioclases with compositions around An_{95} to An_{97} and (2) the larger detrital-plagioclases with compositions of around An_{90} to An_{93} (Albee *et al.*, 1973). Diagenetic overgrowths on detrital grains have compositions similar to the matrix grains. In comparison, large single plagioclase grains in a type 2 breccia have compositions ranging from $An_{95.4}$ to $An_{97.2}$ (Meyer and McCallister, 1973) (Table 5.8).

Orthopyroxene is the other main detrital mineral component in the Cayley lithologies. Analyses from a type 2 breccia suggest a compositional range of $En_{68}Fs_{30}Wo_2$ to $En_{62}Fs_{37}Wo_1$ (Meyer and McCallister, 1973) (Table 5.9).

Most lithologies contain detrital olivine as well as lesser amounts of detrital spinel, metal sulphide and in at least one case zircon.

Lithic Clasts

Apart from reworking clastic-rock fragments only two rock types, both apparently igneous, are at all common: (1) anorthositic clasts and (2) troctolitic clasts. Compositionally the anorthositic clasts in type 2 lithologies appear identical to plagioclase clasts and presumably come from the same source. A mafic mineral, probably orthopyroxene (En_{63}), and a chrome spinel were also encountered (Meyer and McCallister, 1973) (Table 5.10). The orthopyroxene has a markedly higher Ca content than the associated mineral clasts suggesting the possibility of more than one source.

The troctolite fragments are the more common of the two lithic clast types but even so they form less 2 per cent of the rock in the case of the type 2 lithologies. These clasts are fine-grained and consist of laths of plagioclase associated with olivine at times in a subophitic texture. The plagioclases are calcium rich (An_{94}) and compostionally similar to associated single crystal clasts of plagioclase (Tables 5.7 and 5.10). The olivine varies in composition between Fo_{80} and Fo_{90} (Meyer and McCallister, 1973) (Table 5.10). Hodges and Kushiro (1973) described a large spinel troctolite fragment which is almost certainly an isolated clast of the same lithology. As well as plagioclase and olivine, this clast contains subordinate spinel and a fine-crystalline mesostasis. There are actually two varieties of spinel. One

TABLE 5.7.
Composition of phases of a typical type 4 breccia (after Albee *et al.*, 1973).

	Plagioclase	Low-Ca Pyroxene	Olivine	High-Ca Pyroxene	K-rich Interstitial Material	Armalcolite	Ilmenite	Cr-spinel	Rutile	Fe-metal*	Apatite	Whitlockite	Troilite	Bulk Composition Calc. 1990 pts	Bulk Composition LSPET (1973)
vol. %	62.12 ±1.77	20.60 ±1.42	10.20 ±0.72	5.13 ±0.51	0.60 ±0.17	0.55 ±0.17				0.35 ±0.13	0.30 ±0.12		0.15 ±0.09		
wt. %	55.49	23.09	12.92	5.63	0.57			0.86		0.91		0.31	0.22		
						0.215+	0.215+	0.215+	0.215+		0.16+	0.16+			
SiO₂	45.08	52.24	37.17	50.82	65.98	0.35	0.36	0.91	0.31	0.05	0.71	0.59	0.11	45.12	44.65
Al₂O₃	35.37	1.48	0.14	2.82	19.73	1.12	0.14	8.63	0.09	n.a.	0.06	0.64	n.a.	20.28	22.94
Cr₂O₃	n.a.	0.46	0.15	0.77	0.32	1.20	0.31	31.12	0.67	n.a.	n.a.	n.a.	n.a.	0.24	0.10
TiO₂	n.a.	0.74	0.05	2.10	0.32	71.73	55.55	22.38	94.88	n.a.	n.a.	n.a.	n.a.	0.82	0.64
MgO	0.14	26.17	38.82	16.81	0.01	9.91	9.29	8.94	0.07	0.01	0.25	6.12	n.a.	12.15	9.60
FeO	0.44	15.52	23.21	7.42	0.13	13.34	34.01	27.06	0.46	93.73	0.18	1.98	61.86	8.68	7.75
MnO	n.a.	0.26	0.17	0.19	n.a.	0.06	0.33	0.52	<0.01	n.a.	n.a.	n.a.	n.a.	0.09	0.12
CaO	18.74	3.38	0.24	19.53	2.93	n.a.	n.a.	n.a.	n.a.	0.01	53.56	38.03	0.13	12.47	13.34
Na₂O	0.65	0.06	n.a.	0.16	0.16	n.a.	n.a.	n.a.	n.a.	0.01	0.01	0.39	n.a.	0.39	0.39
K₂O	0.13	n.a.	n.a.	n.a.	9.19	n.a.	n.a.	n.a.	n.a.	n.a.	n.a.	0.01	n.a.	0.12	0.11
BaO	0.08	n.a.	n.a.	n.a.	0.40	n.a.	n.a.	n.a.	n.a.	n.a.	n.a.	n.a.	n.a.	0.05	0.02
P₂O₅	n.a.	n.a.	n.a.	n.a.	n.a.	n.a.	n.a.	n.a.	n.a.	0.02	39.52	40.90	0.01	0.13	0.22
ZrO₂	n.a.	n.a.	n.a.	n.a.	n.a.	2.12	0.04	0.02	1.34	n.a.	n.a.	n.a.	n.a.	0.01	0.03
V₂O₃	n.a.	n.a.	n.a.	n.a.	n.a.	0.12	0.02	0.33	0.02	n.a.	n.a.	n.a.	n.a.	<0.01	n.r.
Nb₂O₅	n.a.	n.a.	n.a.	n.a.	n.a.	n.a.	n.a.	n.a.	1.38	n.a.	n.a.	n.a.	n.a.	<0.01	<0.01
Ni	n.a.	n.a.	0.05	n.a.	n.a.	n.a.	n.a.	n.a.	n.a.	6.26	n.a.	n.a.	0.01	0.06	0.02
Co	n.a.	n.a.	n.a.	n.a.	n.a.	n.a.	n.a.	n.a.	n.a.	0.15	n.a.	n.a.	n.a.	<0.01	n.r.
F	n.a.	n.a.	n.a.	n.a.	n.a.	n.a.	n.a.	n.a.	n.a.	n.a.	3.10	0.01	n.a.	<0.01	n.r.
Cl	n.a.	n.a.	n.a.	n.a.	n.a.	n.a.	n.a.	n.a.	n.a.	--	0.74	0.01	n.a.	<0.01	n.r.
S	n.a.	n.a.	n.a.	n.a.	n.a.	n.a.	n.a.	n.a.	n.a.	<0.01	n.a.	n.a.	37.07	0.08	0.12
ΣRE₂O₃ +Y₂O₃	n.a.	n.a.	n.a.	n.a.	n.a.	n.a.	n.a.	n.a.	n.a.	n.a.	2.55	5.93	--	0.01	n.r.
TOTAL	100.63	100.31	100.00	100.62	98.85	99.95	100.05	99.91	99.22	100.21	100.68	94.60	99.19	100.70	100.05
	An 91	En 71	Fo 75	En 49											
	Ab 6	Fs 24	Fa 25	Fs 13											
	Or 1	Wo 5		Wo 38											
	Others 2														

*Elemental abundances. n.a. = Not analized. n.r. = Not reported. + = Values used for calculating bulk composition.

TABLE 5.8.

Representative compositions of detrital plagioclase grains from a type 2 breccia (from Meyer and McCallister, 1973).

Oxides	Feldspar				Recrystallized Feldspar		
SiO$_2$	44.0	44.2	44.0	44.2	44.4	43.9	44.7
TiO$_2$	0.02	0.03	0.02	0.02	0.02	0.02	0.02
Al$_2$O$_3$	35.8	35.9	35.4	36.0	34.7	35.3	34.4
FeO	0.1	0.17	0.13	0.09	0.28	0.19	0.3
MgO	0.08	0.13	0.14	0.08	0.1	0.18	0.18
CaO	19.5	19.1	19.4	19.3	19.2	19.4	19.6
Na$_2$O	0.35	0.47	0.51	0.51	0.31	0.39	0.31
K$_2$O	0.01	<0.01	0.02	<0.01	<0.01	0.01	<0.01
Total	99.9	100.0	99.5	100.1	99.1	99.4	99.8

TABLE 5.9.

Composition of detrital orthopyroxene clasts from a typical type 2 breccia (from Meyer and McCallister, 1973).

Oxides	1	2	3	4
SiO$_2$	52.8	53.5	52.6	53.9
TiO$_2$	0.22	0.23	0.22	0.14
Al$_2$O$_3$	0.67	0.77	0.61	0.67
Cr$_2$O$_3$	0.11	0.15	0.13	0.2
FeO	21.7	21.3	21.9	19.8
MgO	23.0	23.6	23.3	24.9
CaO	0.82	0.65	0.81	0.84
MnO	0.67	0.56	0.54	0.44
Na$_2$O	<0.01	<0.01	<0.01	<0.01
Total	100.0	100.8	100.1	100.8
En	64	66	64	64
Fs	34	33	34	34
Wo	2	1	2	2

TABLE 5.10.

Composition of mineral phases encountered in lithic clasts in a type 2 breccia (from Meyer and McCallister, 1973).

Oxides	Anorthosite Clast			Troctolite Clast			
	Feldspar	Pyroxene	Spinel	Rock	Rock	Feldspar	Olivine
SiO_6	43.9	51.6	—	47.8	44.7	44.6	41.3
TiO_2	0.02	0.09	3.32	0.15	0.83	0.03	0.03
Al_2O_3	35.3	4.63	15.1	24.3	21.2	35.0	0.23
Cr_2O_3	<0.01	0.56	42.7	0.16	<0.16	0.01	0.17
FeO	0.19	19.4	31.0	5.1	8.32	0.23	10.1
MgO	0.18	21.5	6.49	5.62	14.1	0.32	48.6
CaO	19.4	2.4	—	15.5	10.6	19.10	0.35
MnO	—	0.37	—	0.18	0.13	—	0.13
Na_2O	0.39	0.05	—	0.45	0.63	0.63	<0.01
K_2O	0.01	—	—	0.01	0.22	0.06	—
Total	99.4	100.6	98.6	99.3	100.8	99.9	100.9
	An 96.4	Wo 5	Chr 61			An 95.3	Fo 89.6
	Ab 3.5	En 63	Sp 31			Ab 4.5	
	Or 0.1	Fs 32	Usp 8			Or 0.2	

variety is pale yellow to colorless and occurs as small euhedral grains which contains 2-4 mole per cent chromite. The other is a pink spinel which occurs as rounded detrital grains with reaction rims. It is more chromium rich (9-16 mole per cent chromite) than the euhedral variant. The mesostasis consists of plagioclase, olivine, clinopyroxene, ilmenite, metallic iron and troilite set in a dark brown devitrified glass.

Genesis of the Cayley Formation

Contrary to the many pre-Apollo hypotheses there is no evidence to suggest a volcanic origin for the Cayley Formation. The most abundant lithologies returned by Apollo 16 are clastic rocks produced by impact. All of the clastic rocks appear to have had a common source, possibly the type 4 lithologies or an earlier generation of breccias. It is evident from the texture of the type 4 breccias that they have undergone a reasonable degree of thermal metamorphism, at least equivalent to one of the higher metamorphic grades proposed by Warner (1972) for the Fra Mauro lithologies. If type 4 breccias are the primary Cayley lithology they must be the product of a relatively large impact event to have seen the necessary thermal metamorphism. Magnetic properties of the rocks suggest metamorphic temperatures above 770°C, the Curie point of iron (Pearce *et al.*, 1973, Oberbeck *et al.*, 1975).

Any hypothesis concerning the origin of the Cayley Formation must consider these basic observations. Consequently most current hypotheses attempt to connect the Cayley Formation with one of the large basin forming events (Chao *et al.*, 1973; Hodges *et al.*, 1973; Eggleton and Schaber, 1972). Most of the hypotheses consider the Cayley materials to be basin ejecta carried beyond the continuous ejecta blanket, either in ballistic trajectories or as highly-fluidized debris. The distribution of the Cayley Formation around the fringes of the Imbrium ejecta blanket tends to support these arguments. The concept of a highly-fluidized base surge has its merits in that such a phase should develop on the surface of a large debris flow moving in a vacuum (see Chapter 3). Such a flow should also travel faster and further than the main body of the base surge.

Oberbeck *et al.* (1975) have, however, raised at least one very serious objection to the concept of the Cayley Formation being part of the major basin ejecta. They studied the large scale chemistry of the lunar surface as observed by gamma-ray and X-ray spectrometry from orbit. They found that the chemical composition of the Cayley Formation not only varies from locality to locality but that most localities have close chemical affinities to the nearby highland terrain (Fig. 5.24). That is, the Cayley Formation appears to consist of locally-derived materials. This suggests that if the Cayley Formation is related to the large basin forming events, the depositional mechanism must involve the incorporation of large volumes of local materials which does not appear to be consistent with a base surge hypothesis although some local materials could be expected in base surge deposits.

As an alternative, Oberbeck *et al.* (1975) have proposed that Cayley materials are the product of erosion by secondary impacts occuring beyond the continuous ejecta blanket of major basin forming events. Theoretical studies and simulations (see Chapter 3) suggest that secondary impacts produce considerably more than their own mass of debris. Secondary ejecta coupled with landslides triggered by the impact events, transport the locally derived materials to topographic lows to produce smooth-plain units.

A possible alternative is a suggestion proposed by Head (1974). He has argued that the Cayley plains may in part be the original floors of old craters and that the Cayley breccias are fall-back materials and possibly ejecta from

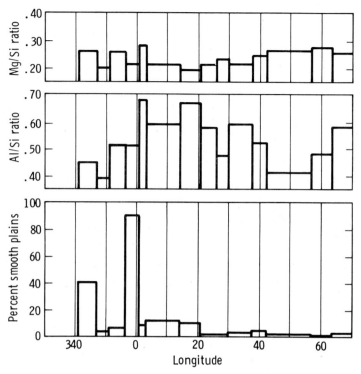

Fig. 5.24. Mg/Si and Al/Si ratios along the Apollo 16 ground track in relation to the areal distribution of the Cayley Formation along the same track (after Oberbeck *et al.*, 1975).

nearby craters. This concept is consistent with the lithologies and the large scale chemistry but does not explain the apparent contemporaneity of the Cayley plains nor their deposition around the large basin ejecta blankets. These relationships may, however, be more apparent than real.

Impact–Induced Fractionation of Lunar Breccias

The intensity of cratering of the lunar highlands attests to the pervasiveness of impact-induced mixing on the lunar surface. Consequently, it is not surprising to find petrographic evidence indicating that the lunar breccias are the product of multiple hypervelocity impacts into complex source regions. The main processes involved during each event are first comminution and then mixing during transport and by secondary impacts. The variety of lithic clasts and the homogeneity of the matrix of any one breccia would suggest that impact mixing has been an extensive and efficient process in the lunar highland. All available data to the present suggest that virtually all of the materials in the upper few kilometers of the lunar crust should be thoroughtly mixed (Head, 1974; Oberbeck, 1971; Oberbeck et al., 1973; Short and Foreman, 1972; Warner et al., 1974). Despite this logic the rocks from any one lunar site are not homogeneous. This may indicate that a differentiation process is occuring during breccia formation and that the mixing process is at least partially reversed (Warner et al., 1975). The two most important chemical processes appear to be crystal fractionation in pools of impact melt and partial melting.

Figure 5.25 presents a schematic model for impact processes as suggested by Warner et al. (1974). However, it may not be necessary to rely as heavily on partial melting as the model appears to suggest. In a single large event such as the Imbrium event there is evidence to suggest that mixing was far from complete and that earlier phases of the base surge (derived from shallower rock units) were compositionally distinct from later phases of the base surge (derived from deeper rock units). Furthermore, the mechanical properties of the different components in the lunar breccias are variable. It thus seems reasonable that some lithologies may crush more readily than others and may be sorted differentially during transport. Wilshire et al. (1973) have explained most of the diversity of the breccias returned by Apollo 16 purely on the basis of differences in the mechanical properties of

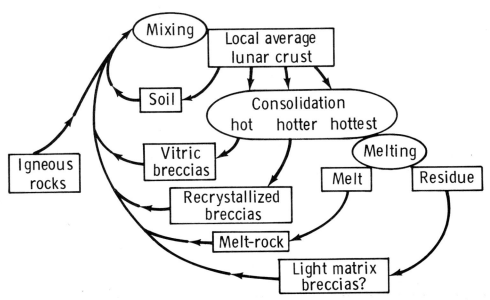

Fig. 5.25. Schematic representation of an impact mixing and partial melting model proposed to explain the diversity of polymict-breccia compositions. Boxes represent materials, ovals represent processes (after Warner et al., 1974, Proc. 5th Lunar Sci. Conf., Suppl. 5, Vol. 1, Pergamon).

the components. Finally, finer-grained materials will tend to melt preferentially and aid partial melting. This is not to imply that the Warner *et al.* (1974) model is incorrect, but other variables need be considered, such as the physical properties of the incorporated materials and the mechanisms of transportation and deposition.

References

Albee, A. L., Gancarz, A. J. and Chodos, A. A., 1973. Metamorphism of Apollo 16 and 17 and Luna 20 metaclastic rocks at about 3.95 AE: Samples 61156, 64423.14-4, 65015, 67483.15-2, 76055, 22006, and 22007. *Proc. Fourth Lunar Sci. Conf., Suppl. 4, Geochim. Cosmochim. Acta, 1:* 569-595.

Anderson, A. T., Braziunas, T. F., Jacoby, J. and Smith, J. V., 1972. Thermal and mechanical history of breccias 14306, 14063, 14270 and 14321. *Proc. Third Lunar Sci. Conf., Suppl. 3, Geochim. Cosmochim. Acta, 1:* 819-835.

Brunfelt, A. O., Heier, K. S., Nilssen, B., Sundroll, B. and Steinnes, E., 1972. Distribution of elements between different phases of Apollo 14 rocks and soils. *Proc. Third Lunar Sci. Conf., Suppl. 3, Geochim. Cosmochim. Acta, 2:* 1133-1147.

Cameron, K. L., and Fisher, G. W., 1975. Olivine-matrix reactions in thermally metamorphosed Apollo 14 breccias. *Earth. Planet. Sci. Lett., 25:* 197-207.

Chao, E. Minkin, J. and Best, J., 1972. Apollo 14 breccias: general characteristics and classification. *Proc. Third Lunar Sci. Conf., Suppl. 3, Geochim. Cosmochim. Acta, 1:* 645-659.

Chao, E. C. T., Soderblom, L. A., Boyce, J. M., Wilhelms, D. E. and Hodges, C. A., 1973. Lunar light plains deposit (Cayley Formation) — A reinterpretation of origin. *Lunar Science—IV,* The Lunar Sci. Inst., Houston, Texas, 127-128.

Clanton, U. S., McKay, D. S., Laughon, R. B. and Ladle, G. H., 1973. Iron crystals in lunar breccias. *Proc. Fourth Lunar Sci. Conf., Suppl. 4, Geochim. Cosmochim. Acta, 1:* 925-931.

Crowell, J. C., 1957. Origin of pebbly mudstones. *Geol. Soc. Amer. Bull., 68:* 993-1009;

Dott, R. H, 1961. Squantum "tillite", Massachusetts — evidence of glaciation or subaqueous mass movement? *Geol. Soc. Amer. Bull., 72:* 1289-1305.

Eggleton, R. E., 1963. Thickness of the Apenninian Series in the Lansberg region of the moon. In: *Astrogeol. Studies Ann. Prog. Rept., August 1961 — August 1962. U. S. Geol. Survey open-file pt. A: U. S. Geol. Survey open-file report,* 19-31.

Eggleton, R. E., 1964. Preliminary geology of the Riphaeus quadrangle of the moon and a definition of the Fra Mauro Formation. In: *Astrogeol. Studies. Ann. Prog. Rept., August 1962-July 1963, pt A: U. S. Geol. Survey open-file report,* 46-63.

Eggleton, R. E. and Marshall. C. H., 1962. Notes on the Apennine Series and pre-Imbrian stratigraphy in the vicinity of Mare Humorum and Mare Nubium. In: *Astrogeol. Studies Semi-Ann. Rept., 1961, U. S. Geol Survey,* 132-137.

Eggleton, R. E. and Schaber, G. G., 1972. Cayley Formation interpreted as basin ejecta. *NASA SP-315,* 29-7 to 29-16.

Elston, D. P., Boudette, E. L. and Schafer, J. P., 1972. Geology of the Apollo 16 landing site area. *U. S. Geol. Survey open-file rept.,* 17 pp.

Engelhardt, W. von, Arndt, J., Stoffler, D. and Schneider, H., 1972. Apollo 14 regolith and fragmental rocks, their compositions and origin by impacts. *Proc. Third Lunar Sci. Conf., Suppl. 3, Geochim. Cosmochim. Acta, 1:* 753-770.

Gilbert, G. K., 1893. The moon's face, a study of the origin of its features. *Philos. Soc. Washington Bull., 12:* 241-292.

Gose, W. A., Pearce, G. W., Strangway, D. W and Larson, E. E., 1972. Magnetic properties of Apollo 14 breccias and their correlations with metamorphism. *Proc. Third Lunar Sci. Conf., Suppl. 3, Geochim. Cosmochim. Acta, 3:* 2387-2395.

Gose, W. A., Pearce, G. W. and Lindsay, J. F., 1975. Magnetic stratigraphy of the Apollo 15 deep drill core. *Proc. Sixth Lunar Sci. Conf., Suppl. 6, Geochim. Cosmochim. Acta* (in press).

Head, J. W., 1974. Stratigraphy of the Descartes region (Apollo 16): Implications for the origin of samples. *The Moon, 11:* 77-99.

Hodges, F. N., and Kushiro, I., 1973. Petrology of Apollo 16 lunar highland rocks. *Proc. Fourth Lunar Sci. Conf., Suppl. 4, Geochim. Cosmochim. Acta, 1:* 1033-1048.

Hodges, C. A., Muehlberger, W. R. and Ulrich, G. E., 1973. Geologic setting of Apollo 16. *Proc. Fourth Lunar Sci. Conf., Suppl. 4, Geochim. Cosmochim. Acta, 1:* 1-25.

Hubbard, N. J., Gast, P. W., Rhodes, J. M., Bansal, B. M., Wiesmann, H. and Church, S. E., 1972. Nonmare basalts: part II. *Proc. Third Lunar Sci. Conf., Suppl. 3, Geochim. Cosmochim. Acta, 2:* 1161-1179.

King, E. A., Butler, J. C. and Carman, M F., 1972. Chondrules in Apollo 14 samples and size analyses of Apollo 14 and 15 fines. *Proc. Third Lunar Sci. Conf., Suppl. 3, Geochim. Cosmochim. Acta, 1:* 673-686.

Kurat, G., Keil, K., Prinz, M. and Nehru, C. E., 1972. Chondrules of lunar origin. *Proc. Third Lunar Sci. Conf., Suppl. 3, Geochim. Cosmochim. Acta, 1:* 707-721.

Lindsay, J. F., 1966. Carboniferous sub-squeous mass-movement in the Manning-Macleay basin, Kempsey, New South Wales. *Jour. Sediment. Pet., 36:* 719-732.

Lindsay, J. F., 1972a. Sedimentology of clastic rocks from the Fra Mauro region of the moon. *Jour. Sediment. Pet., 42:* 19-32.

Lindsay, J. F., 1972b. Sedimentology of clastic rocks returned from the moon by Apollo 15. *Geol. Soc. Amer. Bull., 83:* 2957-2970.

Lindsay, J. F., Summerson, C. H. and Barrett, P. J., 1970. A long-axis clast fabric comparison of the Squantum "tillite", Massachusetts and the Gowganda Formation, Ontario. *Jour. Sediment. Pet., 40:* 475-479.

Lunar Sample Preliminary Research Team, 1971. Preliminary examination of lunar samples from Apollo 14. *Science, 173:* 681-693.

McKay, D. S. Clanton, U. S., Morrison, D. A. and Ladle, G. H., 1972. Vapor phase crystallization in Apollo 14 breccias. *Proc. Third Lunar Sci. Conf., Suppl. 3, Geochim. Cosmochim. Acta, 1:* 739-752.

Meyer, H. O. A. and McCallister, R. H., 1973. Mineralogy and petrology of Apollo 16: rock 60215,13. *Proc. Fourth Lunar Sci. Conf., Suppl. 4, Geochim. Cosmochim. Acta, 1:* 661-665.

Milton, D. J., 1972. Geologic map of the Descartes region of the moon, Apollo 16 premission map. *U. S. Geol. Surv. Misc. Geol. Inv. Map I-748.*

Nelen, J., Noonan, A. and Fredriksson, K., 1972. Lunar glasses, breccias, and chondrules. *Proc. Third Lunar Sci. Conf., Suppl. 3, Geochim. Cosmochim. Acta, 1:* 723-737.

Oberbeck, V. R., 1971. A mechanism for the production of lunar crater rays. *The Moon, 2:* 263-278.

Oberbeck, V R., Horz, F., Morrison, R. H., Quaide, W. L. and Gault, D. E., 1975. On the origin of lunar smooth plains. *The Moon, 12:* 19-54.

Pearce, G. W., Gose, W. A. and Strangway, D. W., 1973. Magnetic studies on Apollo 15 and 16 lunar samples. *Proc. Fourth Lunar Sci. Conf., Suppl. 4, Geochim. Cosmochim. Acta, 3:* 3045-3076.

Pettijohn, F. J., 1957. *Sedimentary Rocks.* Harper and Rowe, N.Y., 2nd ed., 718 pp.

Short, N. M. and Foreman, M. L., 1972. Thickness of impact crater ejecta on the lunar surface. *Modern Geol., 3:* 69-91.

Sutton, R. L., Hait, M. H. and Swann, G. A., 1972. Geology of the Apollo 14 landing site. *Proc. Third Lunar Sci. Conf., Suppl. 3, Geochim. Cosmochim. Acta, 1:* 27-38.

Taylor, S. R., 1975. *Lunar Science: A Post-Apollo View.* Pergamon Press, N. Y., 372 pp.

Trask, N. J. and McCauley, J. F., 1972. Differentiation and volcanism in the lunar highlands: photogeologic evidence and Apollo 16 implications. *Earth Planet. Sci. Lett., 14:* 201-206.

Warner, J. L., 1972. Metamorphism of Apollo 14 breccias. *Proc. Third Lunar Sci. Conf., Suppl. 3, Geochim. Cosmochim. Acta, 1:* 623-643.

Warner, J. L., Simonds, C. H. and Phinney, W. C., 1974. Impact-induced fractionation in the lunar highlands. *Proc. Fifth Lunar Sci. Conf., Suppl. 5, Geochim. Cosmochim. Acta, 1:* 379-397.

Wilhelms. D. E., 1965. Fra Mauro and Cayley Formations in the Mare Vaporum and Julius Caesar quadrangles. *Astrogeol. Studies Ann. Prog. Rept., 1964-1965, Pt. A. U. S. Geol. Survey,* 13-28.

Wilhelms, D. E., 1970. Summary of lunar stratigraphy — telescopic observations. *U. S. Geol. Survey Prof. Pap. 599-F,* 47 pp.

Wilhelms, D. E. and McCauley, J. F., 1971. Geologic map of the near side of the moon. *U. S. Geol. Survey Misc. Geol. Inv. Map. I-703.*

Williams, R. J., 1972. The lithification and metamorphism of lunar breccias. *Earth Planet. Sci. Lett., 16:* 250-256.

Willis, J. P., Erlank, A. J., Gurney, J. J., Theil, R. H. and Ahrens, L. H., 1972. Major, minor, and trace element data for some Apollo 11, 12, 14, and 15 samples. *Proc. Third Lunar Sci. Conf., Suppl. 3, Geochim. Cosmochim. Acta, 2:* 1269-1273.

Wilshire, H. G. and Jackson, E. D., 1972. Petrology and stratigraphy of the Fra Mauro Formation at the Apollo 14 site. *U. S., Geol. Survey Prof. Paper 785,* 26 pp.

Wilshire, H. G., Stuart-Alexander, D. E. and Jackson, E. D., 1973. Apollo 16 rocks: petrology and classification. *Jour. Geophys. Res., 78:* 2379-2392.

Wolfe, E. W. and Bailey, N. G., 1972. Linearment of the Apennine Front — Apollo 15 landing site. *Proc. Third Lunar Sci. Conf., Suppl. 3, Geochim. Cosmochim. Acta, 1:* 15-25.

CHAPTER 6

The Lunar Soil

Introduction

The surface of the moon is blanketed by a thin layer of weakly-cohesive detrital materials which is generally referred to as "soil" or "regolith" (Gault *et al.*, 1966; Rennilson *et al.*, 1966; Christensen *et al.*, 1967; Scott *et al.*, 1967; Shoemaker *et al.*, 1967; Cherkasov *et al.*, 1968; Oberbeck and Quaide, 1967; Quaide and Oberbeck, 1968). The soil is a stratified and continually-evolving sedimentary body produced in large part by hypervelocity meteoroid impact, although locally a pyroclastic volcanic contribution may be present. The understanding of the process by which the lunar soil has accumulated is of particular importance because it may ultimately provide a detailed stratigraphic record of much of lunar history and, in the case of the quiescent post-mare period, it may offer the only stratigraphic record with any continuity. That is, the few meters of lunar soil may offer the only detailed record of in excess of 60 per cent of lunar history. The following chapter is meant to provide some concept of the nature of the lunar soil and an idea of what is currently known of its evolution. The soil evolves in response to poorly understood stochastic processes and as a consequence is difficult to model adequately.

Soil Thickness and Accumulation Rates

Prior to the Apollo missions the thickness of the lunar soil was estimated by studying the geometry of small craters (Oberbeck and Quaide, 1967, 1968; Quaide and Oberbeck, 1968). These estimates proved to be very reliable and still offer the best insight into the variable nature of the soil blanket. Oberbeck and Quaide (1968) found that they could recognize four types of soil thickness distributions, with median thickness values of 3.3, 4.6, 7.5, and 16 meters (Fig. 6.1). The median thickness of the regolith correlates directly with the density of impact craters. The variations in median soil-thickness thus reflect differences in elapsed time since the production of the new rock surface upon which the soil is evolving. The soil begins as a thin deposit of nearly uniform thickness and gradually changes to a thicker deposit with a greater spread of thickness values. That is, the standard deviation of the soil thickness frequency distribution increases with

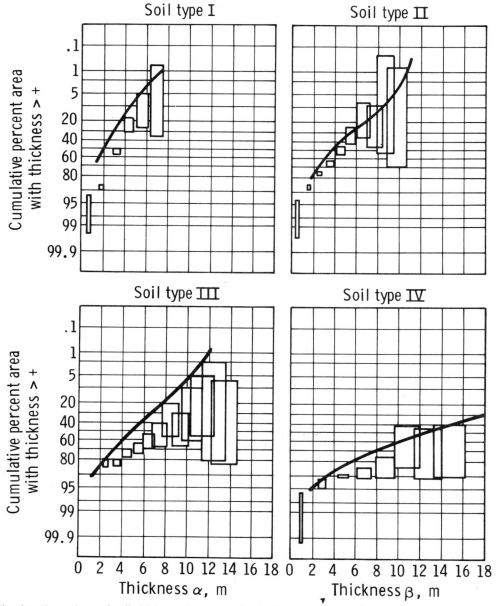

Fig. 6.1. Four observed soil thickness frequency distributions determined from crater morphology (from Oberbeck and Quaide, 1968).

time. The increasing spread of thickness values highlights the stochastic nature of the impact process — soil accumulation is a discontinuous process dependent upon random (in both space and time) hypervelocity impacts.

As a result of the Apollo missions these thickness estimates were, in general, confirmed on the basis of both active and passive seismic experiments (Watkins and Kovach, 1973; Cooper *et al.*, 1974; Nakamura *et al.*, 1975). From these experiments it was concluded that the moon is covered with a layer of low-velocity material that ranges in thickness from between 3.7 and 12.2 m at the Apollo landing sites (Table 6.1). In general the soil thickness increases with the age of the underlying substrate.

Soil accumulation is a self-damping process such that the average accumulation rate decreases with time (Lindsay, 1972, 1975; Quaide and Oberbeck, 1975). If the meteoroid flux had remained constant over time its effectiveness as an agent of erosion would gradually be reduced as the soil blanket grew in thickness. For new material to be excavated from the bedrock beneath the soil an impact must be energetic enough to first penetrate the pre-existing soil layer. As the soil blanket grows, more and more energetic events are required to accomplish the same result. However, returning to Chapter 2, it is evident that the number flux of particles decreases rapidly with increasing particle size and with it the available erosional energy must also decrease. The energy actually available for erosion of bedrock is probably considerably less than 1 per cent of the total meteoritic energy incident on the moon (Lindsay, 1975) (Fig. 6.2). Whatever the history of bombardment there should be rapid initial accumulation of soil followed by gradually decreasing growth rates.

Modelling of such complex processes requires more knowledge than is currently available about the meteoroid flux, and therefore some simplifying

TABLE 6.1

Estimates of regolith thickness based on seismic data from each of the Apollo landing sites (data from Nakamura *et al.*, 1975).

Apollo Station	Thickness (m)	Minimum Age of Substrate (x 10^9 yr)
11	4.4	3.51
12	3.7	3.16
14	8.5	3.95
15	4.4	3.26
16	12.2	—
17	4.0, 8.5	3.74

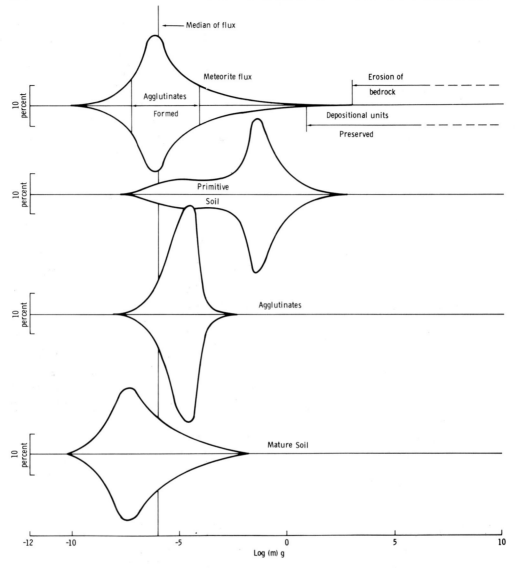

Fig. 6.2. The meteoroid flux, a primitive and mature soil and an agglutinate distribution expressed as weight per cent per decade of mass. Note the similar shape of the meteoroid flux distribution and the lunar soil (from Lindsay, 1975).

assumptions are necessary. Quaide and Oberbeck (1975) have used a Monte Carlo approach to study average cases of soil growth on mare surfaces. If the meteoroid flux has been constant since the flooding of the mare the thickness (Th) of the soil blanket is related to its age (A) by

$$Th = 2.08 \, A^{0.64} \qquad\qquad (6.1)$$

where Th and A are in meters and aeons respectively. The accumulation rate (dTh/dA) decreases exponentially with time (Fig. 6.3). This observation has considerable bearing on the understanding of the texture and mineralogy of the soil. As the accumulation rate declines more and more of the kinetic energy of the meteoroid flux is redirected into mixing and reworking the soil. This extra energy is particularly important in modifying the grain-size parameters of the soil which in turn affects the nature of the impact-produced glass particles and the glass content of the soil. Because smaller volumes of fresh bedrock material are added with time, any materials that are cumulative will increase in concentration — this is true of impact glasses (Lindsay, 1971a, 1972c) and should be true of the meteoritic materials themselves.

Soil Stratigraphy and Dynamics

The lunar soil is layered. This observation was first made in drive-tube core samples returned by Apollo 12 (Lindsay et al., 1971) and has since been confirmed in deeper drill cores collected by subsequent Apollo missions (Heiken et al., 1973; Duke and Nagle, 1975) (Fig. 6.4).

The Apollo 15 deep drill core intersected 42 major stratigraphic units and a large number of sub-units ranging from 0.5 cm to 20 cm in thickness (Heiken et al., 1973). The thickness-frequency distribution of the units is

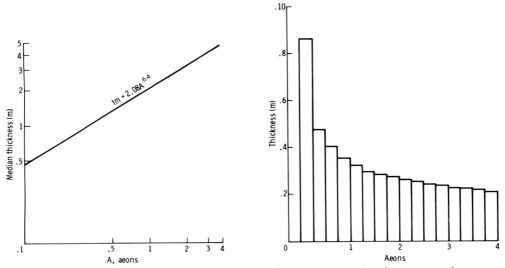

Fig. 6.3. Accumulation rate and thickness of lunar soil for a constant cratering rate over 4 aeons. The accumulation rate is damped exponentially by the gradually thickening soil blanket (after Quaide and Oberbeck, 1975).

Fig. 6.4. Sampling the lunar soil with a drive tube during the Apollo 12 mission (NASA Photo AS12-49-7286).

bimodal with the strongest mode at 1.0 to 1.5 cm and a secondary mode at 4.5 to 5.0 cm (Fig. 6.5) (Lindsay, 1974). Preliminary analysis of drill-core samples from the Apollo 16 landing site suggests that this distribution may be typical of the lunar soil (Duke and Nagle, 1975).

The units are distinguished on the basis of soil color and texture, particularly variations in grain size. The contacts of the units are generally sharp, although at times difficult to observe because the textural differences are subtle. Some units have transitional contacts and structures similar to flame structures were observed at the base of other units. Both normal and reverse graded beds have been described (Lindsay *et al.,* 1971; Heiken *et al.,* 1972; Duke and Nagle, 1975).

The presence of normal and reverse graded bedding in the lunar soil (Lindsay *et al.,* 1975; Heiken *et al.,* 1972; Duke and Nagle, 1975) suggests that a single set of depositional processes were operative during the formation of at least some stratigraphic units. Grain size data from individual units suggests that two processes may be active in the transport of soil materials following a hypervelocity impact, i.e. base surge and grain flow (Lindsay, 1974). Both of these processes are discussed in detail in Chapter 3.

An impacting meteoroid, by virtue of the heat energy generated, may produce large volumes of gas and a base surge may develop. Detrital particles transported in a base surge are fluidized by the upward flow of escaping gases. The settling velocity (under Stokes Law) of the particles is determined by their size and to some extent their shape. Consequently, larger and more spherical particles tend to move towards the base of the fluidized flow and produce a normally-graded detrital mass. However, the supply of fluidizing gases is not necessarily in any direct proportion to the size or velocity of the

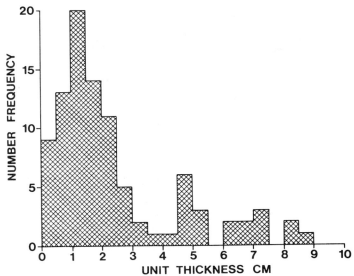

Fig. 6.5. Thickness frequency distribution of soil units intersected by a deep drill core at the Apollo 15 site. The distribution is unexpectedly bimodal (after Lindsay, 1974).

flowing mass, with the result that one of two things can happen. If the gas supply is abundant fluidization may continue until the flow loses momentum and comes to rest as a normally-graded depositional unit. If on the other hand, the gas supply is limited fluidization may cease before the flow has completely come to rest. In this case inertial grain flow will dominate. During inertial grain flow a dispersive pressure is generated which causes larger particles to migrate upwards through the flow to the region of least shear strain resulting in a reverse-graded bed. Some depositional units show the effects of both in that larger particles concentrate in the middle of the unit (Lindsay, 1974) (Fig. 6.6). Similar sorting effects are also apparent in particle shape studies where more spherical particles are found to congregate in the middle of the depositional unit (Lindsay, 1974) (Fig. 6.7).

While it is reasonable to suggest that at least some strata in the lunar soil are formed by single impact events, it is by no means clear that all strata have such simple origins (Fig. 6.8). Hypervelocity cratering experiments suggest that an inverted stratigraphy may be preserved in the ejecta blankets of successive cratering events (see Chapter 3 for details). If so, the lunar soil may be a complex of first generation units interbedded with inverted sequences of older generation depositional units. Evidence of inverted sequences is limited. Recently however, Gose *et al.* (1975) have found systematic changes in the magnetic properties of the Apollo 15 core which they interpret as evidence of stratigraphic inversions. The evidence is however ambiguous and the deviations may be explained by other causes.

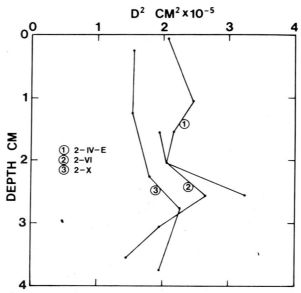

Fig. 6.6. Mean grain size (D) squared (cm) as a function of depth in three lunar soil layers from the Apollo 15 deep drill core (after Lindsay, 1974). Larger particles tend to concentrate in the middle of the units.

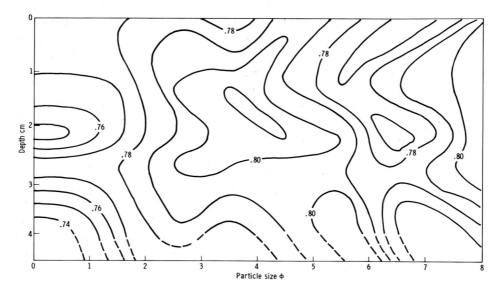

Fig. 6.7. Two dimensional sphericity as a function of grain size and depth in a depositional soil unit from the Apollo 15 deep drill core. The more spherical particles tend to concentrate in the middle of the unit (after Lindsay, 1974).

Fig. 6.8. A small crater in lunar soil surrounded by a thin ejecta blanket. The crater has not penetrated the soil blanket but has simply redistributed the soil to produce a new stratigraphic unit which may or may not survive subsequent reworking (NASA Photo AS15-85-11466).

Whether or not stratigraphic inversions exist, there is clear evidence that some depositional units are reworked by micrometeoroids. For example, if each unit is a depositional unit we could expect the thickness-frequency distribution in Figure 6.5 to reflect, in some simple way, the meteoroid flux. That is, the number of units should increase exponentially as the unit thickness decreases. However, in both cores so far studied this is not the case, the number frequency peaks then declines sharply (Lindsay, 1974; Duke and Nagle, 1975). Very thin depositional units are thus either not formed or are destroyed by reworking. In most of the lunar cores evidence of such reworking can be observed directly. For example, some strata consist of dark soil with rounded clods of lighter soil included within them, somewhat similar to a mudchip breccia. Duke and Nagle (1975) call this texture an "interclastic soil chip breccia." The clods or soil chips are no more cohesive than the matrix soil. Since the clods occur only near the upper contact of the unit it seems most likely that we are looking at an incompletely homogenized compound unit. A thin layer of light soil apparently overlay a thicker dark soil unit, and the clods are the incompletely mixed remnants of the thin layer of light soil. Undoubtably, many other thin units of soil have been reworked in the same way but are not visible because thay lack distinctive physical properties or were completely homogenized. Many fresh glass particles show a selenopetal (lunar geopetal) orientation, related to depositional configuration (Duke and Nagle, 1975). These glass splashes struck the lunar surface while hot, outgassed and vesiculated on their upper surfaces and incorporated dust into their bases. Such glass particles in many units indicate that they are still undisturbed and in their depositional configuration. In some units however, the particles are overturned and fractured indicating reworking. There is no evidence of overturned sequences from selenopetal structures.

Probabilistic models for mixing and turnover rates due to hypervelocity impact suggest that the upper millimeter of the lunar soil is the primary mixing zone (Fig. 6.9) (Gault *et al.*, 1974). This mixing zone is probably coincident with the zone in which agglutinates form. Agglutinates are complex constructional glass particles and are discussed in a following section (p. 261). Below this mixing zone the rate of turnover and mixing decreases very rapidly with increasing depth which explains the well-preserved soil stratigraphy.

Soil Density

The mean density of the particulate material forming the lunar soil ranges from 2.90 to 3.24 g cm^{-3} depending upon the nature of source materials. Basaltic mare soils tend to be denser than the more anorthositic highland soils. It has been found experimentally that the porosity of lunar soils may range from 41 to 70 per cent with the result that soil density is variable (Carrier *et al.*, 1974). The main reason for the range in porosities is probably connected with the abundance of extremely-irregularly-shaped

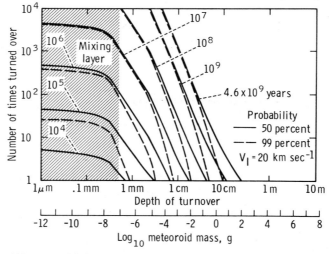

Fig. 6.9. A probabilistic model for soil reworking based on a constant meteoroid mass flux. The number of times the soil is turned over is shown as a function of depth and time. Most reworking occurs in a millimeter surface zone where agglutinates are formed (from Gault *et al.*, 1974; *Proc. 5th Lunar Sci Conf., Suppl. 5, Vol. 3*, Pergamon).

glass particles (agglutinates) which are discussed in a following section. Data from core tubes driven into the lunar surface have shown that soil density increases in a logarithmic manner with depth (Carrier *et al.*, 1974). The density profile can be approximated by

$$\rho = \rho_o + 0.121 \, \ln (Z + 1) \tag{6.2}$$

where ρ_o and ρ are the density at the surface and at some depth Z (cm). ρ_o is approximately 1.38 g cm^{-3} at the Apollo 15 site. The continual reworking of the lunar surface apparently keeps the surface layers of the soil loose but at depth the vibration due to the passage of numerous shock waves causes the soil to increase in density.

Composition of Lunar Soils

The chemistry and mineralogy of the lunar soils, for the most part, reflect the composition of the underlying bedrock (Table 6.2). Thus soils from mare areas have an overall basaltic composition with a high Fe content, while the highland soils tend to be more anorthositic in composition and

TABLE 6.2

Major element chemistry of average Apollo 15 soils and basalts. There is a general similarity between the soils and basalts but the higher Al_2O_3 value of the soils indicates the addition of some highland materials (after Taylor, 1975).

	Average Soil	Olivine Basalt	Quartz Basalt
SiO_2	46.61	44.2	48.8
TiO_2	1.36	2.26	1.46
Al_2O_3	17.18	8.48	9.30
FeO	11.62	22.5	18.6
MgO	10.46	11.2	9.46
CaO	11.64	9.45	10.8
Na_2O	0.46	0.24	0.26
K_2O	0.20	0.03	0.03
P_2O_5	0.19	0.06	0.03
MnO	0.16	0.29	0.27
Cr_2O_3	0.25	0.70	0.66
TOTAL	100.13	99.46	99.08

have high Al and Ca values (Table 6.3). It has been estimated that, in general, 95 per cent of a soil sample at any given point on the moon is derived from within 100 km (Fig. 6.10) (Shoemaker *et al.*, 1970). The proportion of exotic components decreases exponentially with distance. Orbital chemical data further emphasize the local derivation of the bulk of the soil. Secondary X-rays generated at the lunar surface by primary solar X-rays allow a determination of Al, Mg, and Si (Adler *et al.*, 1972a, 1972b). The secondary X-rays are generated in the 10 μm surface layer and thus offer a good picture of the lateral distribution of the lunar soil with a spatial resolution of about 50 km. In Figure 6.11 it can be seen that in general high Al/Si values coincide with highland areas and low values coincide with the mare areas. The boundaries between the two are sharp at the available resolution. Lateral movement of detrital material thus cannot be rapid despite continued reworking for long time periods by the meteoroid flux.

Soil Petrography

The lunar soil consists of three basic components: (1) rock fragments, (2) mineral grains and (3) glass particles. The composition of these three basic components varies considerably from one site to another, depending upon the nature of the bedrock. They also vary in abundance laterally and with depth in the soil in response to the addition of more distant exotic components, the degree to which the soil has been reworked (its age), and in response to bedrock inhomogeneities (Table 6.4).

TABLE 6.3

Average major element chemistry for highland and mare soils (after Turkevich *et al.*, 1973).

Element	Per cent of atoms			Weight per cent of oxides		
	Mare	Highland	Average surface	Mare	Highland	Average surface
O	60.3 ± 0.4	61.1 ± 0.9	60.9			
Na	0.4 ± 0.1	0.4 ± 0.1	0.4	0.6	0.6	0.6
Mg	5.1 ± 1.1	4.0 ± 1.1	4.2	9.2	7.5	7.8
Al	6.5 ± 0.6	10.1 ± 0.9	9.4	14.9	24.0	22.2
Si	16.9 ± 1.0	16.3 ± 1.0	16.4	45.4	45.5	45.5
Ca	4.7 ± 0.4	6.1 ± 0.6	5.8	11.8	15.9	15.0
Ti	1.1 ± 0.6	0.15 ± 0.08	0.3	3.9	0.6	1.3
Fe	4.4 ± 0.7	1.8 ± 0.3	2.3	14.1	5.9	7.5

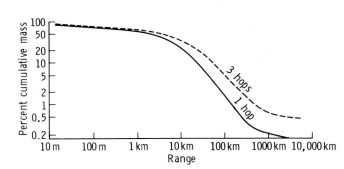

Fig. 6.10. Dispersion of detrital soil materials by impact cratering. Most soil materials are locally derived (after Shoemaker *et al.*, 1970; *Proc. Apollo 11 Lunar Sci. Conf., Suppl. 1, Vol. 3.*, Pergamon).

Lithic Clasts

Lithic clasts are the dominant component in particle size ranges larger than 1 mm (60 to 70 per cent of the 1-2 mm size range). Below 1 mm their importance decreases rapidly as they are disaggregated to form mineral and glass particles. The coarser the mean grain size of the soil the greater, in general, the abundance of lithic fragments.

Three categories of clastic fragments should be considered (1) igneous rock fragments, (2) crystalline breccias and (3) soil breccias.

Igneous Rock Clasts

Mare basalts with the full range of textures are by far the most common igneous lithic fragment and dominate the soils at the mare sites. At the highland sites some igneous rock fragments are encountered but most of them appear to be reworked clasts from the underlying crystalline breccia bedrock. Most of these clasts consist of plagioclase grains and come from grabbroic-noritic-anorthositic sources or, less frequently, from a highland basalt source. Small numbers of feldspathic-lithic fragments occur in the mare soils, providing the first indication of the composition of the lunar crust (Wood *et al.*, 1970). Because the grain size of the minerals is large compared to the size of the clasts, positive identification of many lithologies is difficult. More complete information is available from the chemistry of homogeneous impact melt glasses which is discussed in a following section.

Crystalline Breccia Fragments

Crystalline breccia fragments dominate the lithic clasts in all of the highland soils. They are bedrock fragments derived from units such as the Fra Mauro and Cayley Formations. Fragments of all of the matrix types are represented but again, because of the size of the clasts in relationship to the grain size of the detrital materials forming the breccias, it is not possible to

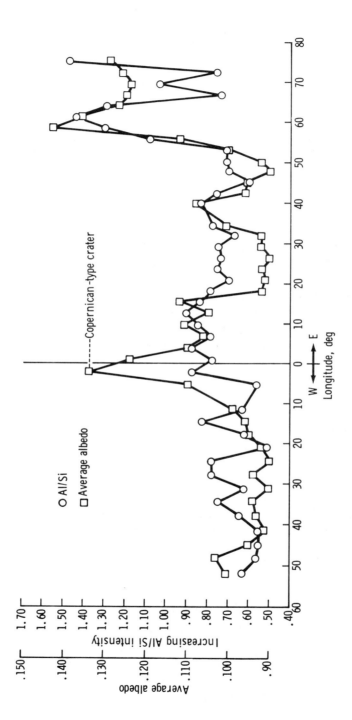

Fig. 6.11. Comparison of Al/Si ratios from the orbital x-ray fluorescence experiment with albedo for a single revolution of the moon. Note that in general the two variables are sympathetic and the highland mare transition is very abrupt indicating little lateral dispersal of soil (from Adler *et al.*, 1972b).

TABLE 6.4

Comparison of the modes of the 90 to 150 μm fraction of a range of soils from the Apollo 14 site. Coarse grained soils contain few agglutinates (adapted from McKay et al., 1972).

Components	14259	14148	14156	14149	14230, 113	14230, 121	14230, 130	14141
	Surface comprehensive samples	Trench top (Sta. G)	Trench middle (Sta. G)	Trench bottom (Sta. G)	Core Sta. G (-8.0 cm)	Core Sta. G (-13.0 cm)	Core Sta. G (-17.0 cm)	Cone Crater surface
Agglutinates	51.7	50.2	47.7	26.4	53.2	57.0	51.5	5.2
Breccias								
Recrystallized	20.3	24.2	23.4	27.0	13.5	16.3	27.0	49.5
Vitric	5.0	4.6	–	8.2	–	4.3	0.4	7.8
Angular Glass Fragments								
Brown	6.7	6.0	7.8	6.0	8.4	7.3	5.2	7.6
Colorless	3.3	2.4	3.2	2.8	1.9	3.0	0.4	2.6
Pale Green	–	0.8	0.9	–	–	1.3	0.9	–
Glass Droplets								
Brown	1.3	1.8	3.2	1.6	4.8	0.3	2.6	2.0
Colorless	0.6	–	0.4	2.0	–	–	–	0.2
Pale Green	0.3	0.2	1.5	0.4	–	–	–	–
Clinopyroxene	3.7	1.8	4.3	7.4	6.8	3.0	3.0	8.0
Orthopyroxene	0.6	1.4	1.3	3.6	0.6	0.6	–	3.8
Plagioclase	4.7	3.0	5.4	7.8	6.5	3.3	3.4	7.6
Olivine	–	0.8	0.2	–	0.6	–	0.4	0.4
Opaque Minerals	–	0.4	0.4	0.4	–	–	–	0.4
Ropy Glasses	0.6	0.2	0.2	–	2.9	1.7	2.1	–
Basalt	1.0	2.0	–	6.4	0.3	1.6	2.1	4.2
No. of Grains Counted	300	500	500	500	300	300	234	500

evaluate the relative abundance of different lithologic types.

Soil Breccias

Soil breccias are impact-lithified breccias formed directly from the lunar soil. The coarser grain size fractions of all lunar soils include a large proportion of soil breccia fragments. The soil breccias are particularly important in the evolution of the soil because like agglutinates they are constructional particles.

Soil breccias are poorly sorted agglomerations of rock, mineral and glass fragments which to a large extent preserve the texture of the parent soil. They are characterized by a very open discontinuous framework and are estimated to have a void space of about 35 per cent. The breccias have densities of around 2 g cm⁻³ which is only slightly greater than unconsolidated soil (Waters et al., 1971; McKay et al., 1970). The spaces between the larger clasts of the framework are filled with glass-rich clastic materials with an average grain size of about 50 μm (i.e. close to the mean grain size of typical lunar soils). The glass forming this matrix tends to be well sorted, closely fitted and plastically molded against the larger clastic rock fragments. In a general way the texture of these rocks resembles terrestrial ignimbrites (Waters *et al.*, 1971), but contrasts sharply with the texture of shock-lithified materials from chemical explosions (McKay *et al.*, 1970). Glass fragments exhibit a wide variety of devitrification features, ranging from incipient to complete devitrification.

There are two prominent textural features of the soil breccias which provide insights into their despositional history: (1) layering and (2) accretionary lapilli. Very-weakly-defined layering or bedding is seen in many of the soil breccias, generally in the same samples where accretionary lapilli are abundant (Waters *et al.*, 1971; Lindsay, 1972). Accretionary lapilli are present in soil breccias from all of the Apollo landing sites and appear to be an intrinsic characteristic of the lithology (McKay *et al.*, 1970, 1971; Lindsay, 1972a, 1972b). The accretionary structures range in size from 50 μm to 4mm and at best are weakly defined (Fig. 6.12). In some samples they form complex aggregates of several lapilli. Some aggregates have an outer accretionary layer surrounding the entire structure. The lapilli generally have a core of one or more larger detrital grains which appear to have acted as a nucleus as the structure accreted. The core is surrounded by alternating layers of dark fine-grained glassy material and of larger detrital grains. The shape of the accretionary layers changed as the lapillus developed. Layers may pinch out or the shape may be changed suddenly by the incorporation of a large detrital grain or another lapillus. Most lapilli are roughly circular or ovoid in section. They are generally similar to terrestrial volcanic accretionary lapilli except that the layering is much more weakly defined (Moore and Peck, 1962).

Genesis of Soil Breccias. The texture of the soil breccias suggests that they have originated, not by direct shock lithification due to meteoroid impact, but in a base surge generated by such an impact. The accretionary lapilli may have formed in turbulent areas of the hot base surge or in the impact cloud which later forms the fall back (McKay *et al.*, 1970). The low density of the breccias and the draping of matrix glass around larger clasts suggests that lithification results from sintering. At the time of sintering the breccias were heated to the yield temperature of the finest glass fraction. Without an atmosphere sintering proceeds at a much lower temperature than in the terrestrial environment and in the absence of pressure the nature of breccias formed by viscous sintering becomes very time and temperature dependent (Uhlmann *et al.*, 1975; Simonds, 1973). An analysis of the kinetics of sintering has delineated several regions in which breccia types should form (Uhlmann *et al.*, 1975) (Fig. 6.13). In the first region amorphous matrix breccias form in very short time intervals. Time is sufficient for matrix

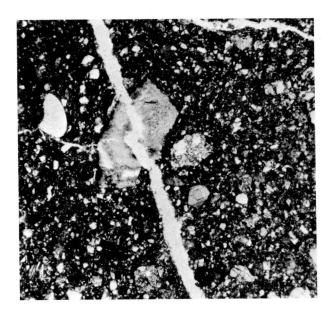

Fig. 6.12. An accretionary lapillus in a soil breccia from the Apollo 14 landing site. Larger detrital grains form the core of the lapillus. The lapillus is approximately 1 mm in diameter (NASA Photo S-71-30835).

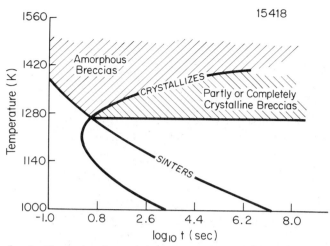

Fig. 6.13. Curve for significant sintering and time-temperature-transformation curve corresponding to a just-detectable degree of crystallinity for Lunar Composition 15418. Regions indicate where amorphous breccias and breccias with detectable crystallinities are formed (from Uhlmann *et al.*, 1975).

sintering to occur but not sufficient for any significant crystallinity to develop. In the second region partly or completely crystalline breccias develop due to the increased time available. In the third region (unshaded in Fig. 6.13) breccias do not form because there is insufficient time for sintering to occur or the materials will crystallize before significant sintering can occur.

Maturity of Soil Breccias. The mineralogic maturity of a sedimentary rock is a measure of the total energy involved in its formation. The main energy source on the lunar surface is the meteoroid flux. In the case of the lunar soil energy from successive impacts is released through (1) vitrification and (2) comminution. A mature soil breccia should, therefore, contain an abundance of glass and an abundance of fine-grained detrital material most of which is glass (i.e. matrix material). The proportion of glass plus matrix material can thus be used as a practical measure of maturity for soil breccias.

Maturity is probably most readily visualized in the form of a ternary diagram with end members of (1) glass plus matrix (defined for practical purposes here as <15 μm) (2) mineral fragments and (3) lithic clasts (Fig. 6.14). Most of the samples cluster relatively tightly near the glass plus matrix apex indicating generally very-mature detrital materials. The soil breccias thus probably formed from soils which had evolved in a regular manner with a few introductions of coarse impact-derived bedrock materials. It also suggests that the events which formed the breccias were probably not large enough themselves to excavate bedrock but simply sintered existing soil. Only five samples lie beyond the main cluster and closer to the lithic apex.

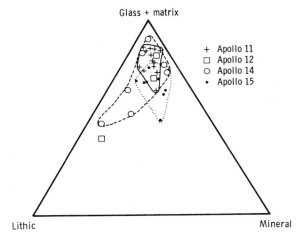

Fig. 6.14. Ternary diagram showing the relative maturity of soil breccias from several Apollo landing sites. Apollo 11 field, solid line; Apollo 14 field, dashed line; and Apollo 15 field, dotted line (after Lindsay, 1972b).

The anomalous Apollo 14 sample can be associated with coarse soils excavated from Cone Crater, and the anomalous Apollo 15 breccia may come from Elbow Crater.

Grain Size of Soil Breccias. Soil breccias are relatively fine grained which, combined with the difficulty of identifying small sintered glass particles and agglutinates, makes grain size determinations difficult. Lindsay (1972a, 1972b) attemped to determine the grain-size distribution of soil breccias using thin section data and Friedman's (1958; 1962) conversions to determine sieve size equivalents. In general the mean grain sizes determined in this way are consistently much finer than equivalent soils from the same site (Table 6.5). This suggests two possibilites: (1) it is not possible to identify agglutinates in thin sections and they are being analysed in terms of their detrital components or (2) because soil breccias are formed from the soil by larger impact events many or most agglutinates are destroyed during crater

TABLE 6.5

Grain size of lunar soil breccias (from Lindsay, 1972b).

Mission	Mean Median		Number of Samples
	ϕ	μm	
11	5.99 $\pm$ .61	15.7	14
12	4.20	54.4	3
14	4.75 $\pm$ 1.76	37.2	7
15	5.14 $\pm$ 1.00	28.4	11

excavation. Detailed thin section analysis of the breccias combined with skewness data from the soils suggests that many agglutinates are destroyed by larger layer-forming impact events.

Regardless of the true grain size of the soil breccias the measured grain size varies sympathetically with the grain size of the local soils. This effect is particularly noticeable in the vicinity of larger craters which penetrate to bedrock and produce coarser soil breccias at the Apollo 14 site all come from the vicinity of Cone Crater. Similarly at the Apollo 15 site the coarse-grained breccias come from the rim of Dune Crater and from close to Elbow Crater both of which penetrate the soil blanket. As with the petrographic data this again emphasises the extremely local nature of the soil breccias.

Roundness of Detrital Grains in Soil Breccias. Communition is the main process by which the morphology of detrital soil particles is modified. As a consequence mineral grains incorporated into the soil breccias are generally very angular (Table 6.6). With the exception of breccia samples from the Apollo 14 site, plagioclase and pyroxene grains have roundness value of between approximately 0.1 and 0.2 resulting in small roundness sorting values. These values are almost certainly determined by the mechanical properties of the minerals themselves.

The effects of communition on roundness are best illustrated by the Apollo 14 soil breccias. The clastic rocks of the Fra Mauro Formation contain moderately-well-rounded plagioclase grains (up to 0.6, see Chapter 5). The soils and soil breccias are derived from the Fra Mauro Formation and inherit many of these comparatively (by lunar standards) well-rounded grains (Fig. 6.15). Continued reworking of the soils by the micrometeoroid flux gradually reduces them to smaller angular grains. Consequently, when the mean roundness of plagioclase grains is plotted as a function of the maturity of the breccia, a strong negative linear dependence can be seen (Fig. 6.16). That is, as more energy is applied to the reworking of the original soil, the

TABLE 6.6

Mean roundness values for plagioclase grains from lunar soil breccias. A minimum of 50 grains (100 to 200 μm in diameter) were measured for each sample (data from Lindsay, 1972b).

Mission	Grand Mean Roundness	Grand Mean Roundness Sorting	Number of Samples
11	0.197 ± .031	0.100 ± .015	14
12	0.243	0.135	3
14	0.242 ± .055	0.114 ± .008	7
15	0.175 ± .021	0.090 ± .019	11

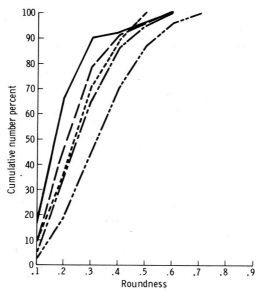

Fig. 6.15. Cumulative roundness—frequency distribution for plagioclase grains from Apollo 14 soil breccias (after Lindsay, 1972a).

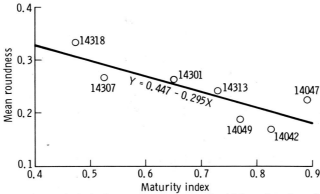

Fig. 6.16. Mean roundness of plagioclase grains from Apollo 14 breccias plotted as a function of maturity. Mean roundness decreases with increasing maturity due to comminution of detrital grains with large inherited roundness values (after Lindsay, 1972b).

mean roundness of the plagioclase grains is reduced by comminution (Lindsay, 1972a, 1972b).

Mineral Grains in Soils

Mineral grains are the dominant detrital particles in the intermediate grain sizes, particularly between 3ϕ (62.5 μm) and 4ϕ (32.1 μm) on the coarse side of the mean grain-size of the bulk soil (Table 6.7). This distribution probably reflects the grain size of the source rocks to a large extent although the physical properties of the minerals themselves are also important.

As is the case for lithic fragments the detrital mineral grains present in a soil reflect, for the most part, the nature of the underlying bedrock. The grains are generally highly angular except at the Apollo 14 site where moderate degrees of roundness are inherited from the source materials. Most mineral grains, but particularly the plagioclases, show much evidence of

TABLE 6.7

Modal composition of two highland soils from the Apollo 14 site. Sample 14141 is a coarse grained texturally immature soil with finer agglutinates whereas sample 14003 is fine grained, texturally mature and contains an abundance of agglutinates (adapted from McKay *et al.*, 1972).

Components	14141,30				14003,28			
	150-250μ	90-150μ	60-75μ	20-30μ	150-250μ	90-150μ	60-75μ	20-30μ
Agglutinates	5.3	5.2	6.5	12.5	54.2	60.3	56.5	43.5
Microbreccias								
Recrystallized	57.5	47.2	36.5	15.5	19.2	20.5	16.5	7.0
Vitric	6.9	6.8	16.5	5.5	4.4	3.0	1.0	0.5
Angular Glass Fragments								
Brown	2.3	6.2	4.0	7.0	7.2	4.3	8.0	6.5
Colorless	0.3	2.6	1.5	5.0	2.0	3.0	3.0	3.5
Pale Green	0.3	—	—	—	0.2	—	—	—
Glass Droplets								
Brown	0.8	2.0	0.5	2.0	2.2	0.3	2.0	7.0
Colorless	—	0.2	—	—	0.2	0.3	—	—
Ropy Glasses	—	—	—	—	—	1.0	0.5	—
Clinopyroxenes	4.1	8.0	15.5	18.5	3.0	2.3	5.5	18.5
Orthopyroxenes	1.2	2.0	2.5	9.5	1.2	1.3	4.5	5.0
Plagioclase	5.6	6.6	11.5	21.5	3.6	2.3	1.5	7.0
Olivine	0.5	0.4	1.5	0.5	1.0	—	—	—
Opaque Minerals	—	—	—	—	—	—	—	1.5
Basalt	5.9	4.2	0.5	—	0.8	1.3	—	—
Anorthosite	0.3	1.2	—	—	0.2	—	—	—
Tachylite	8.9	7.0	3.0	2.5	—	—	1.0	—
No. of Grains counted	400	500	200	200	500	300	200	200

shock modification as a consequence of the impact environment. Many plagioclase crystals have been disrupted by shock to such an extent that they have been converted to diaplectic glasses.

Plagioclase is the ubiquitous mineral in the lunar soils. The anorthite content is generally high in any soil although the range of values in mare soils is generally greater (An_{60} to An_{98}) than in highland soils (generally An_{90} or higher); highland plagioclases, however, may also have a large compositional range (Fig. 6.17) (Apollo Soil Survey, 1974).

A small number of K-feldspar grains have been encountered in the lunar soils. Most are small and attached to, or intergrown with, plagioclase. They generally have an Or range of from 70 to 86 mole per cent.

Pyroxenes are present in most soils and like the plagioclase frequently show signs of shock modification. In mare soils they are almost exclusively clinopyroxene (augite) derived from the mare basalts. Highland soils contain a relatively large proportion (over a third) of orthopyroxene as well as clinopyroxene. The pyroxenes in the highland soils probably were mainly derived from norite and highland basalt clasts in the crystalline breccia basement rock, although pyroxenes are common as mineral clasts in many highland breccias. The orthopyroxenes are mostly bronzites but range from En_{85} to En_{61}. In Fig. 6.18 detrital pyroxenes from highland soils (Apollo 14) are contrasted with pyroxenes from the mare sites (Apollos 11 and 12). Augites and subcalcic augites dominate the mare pyroxenes whereas the highland pyroxenes extend into the low-calcium range (Apollo Soil Survey, 1974). Mare pyroxenes are also distinguished from their highland equivalents by the presence of extreme compositional zoning.

The pyroxene/plagioclase ratio of most texturally mature soils consistently increases with decreasing grain size (Table 6.7; see sample 14003). This may simply reflect the differing mechanical properties of the two minerals or, as Finkelman (1973) has suggested, it may be due to fine grain sizes being transported over greater distances. However, the net result is an increase in mafic elements in finer grain sizes which has considerable bearing on the chemistry of some glass particles (particularly agglutinates which are discussed in a following section).

Olivine is present in most soils but is very variable in proportion. Only some mare basalts contain a large percentage of olivine so there is a tendency for it to vary from sample to sample at any one mare site. Olivine is generally only present in small amounts (7 per cent) in highland soils, where it is probably derived from pre-existing troctolite fragments and perhaps from a "dunite" source. The olivines mostly lie in range of Fo_{75} to Fo_{60} with a peak around Fo_{67} (Fig. 6.19). A few grains are either extremely magnesian (Fo_{89}) or extremely iron-rich (Fo_{14}) (Apollo Soil Survey, 1974). In general

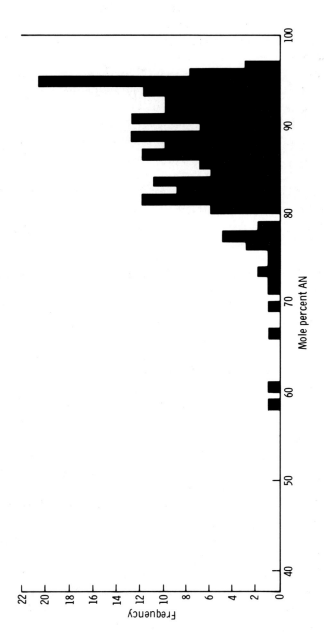

Fig. 6.17. Histogram of plagioclase from highland soils collected by Apollo 14 (from Apollo Soil Survey, 1974).

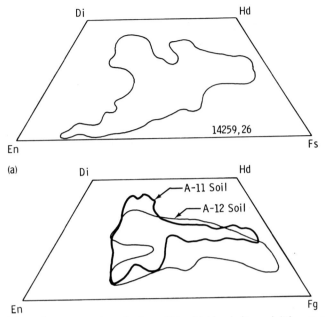

Fig. 6.18. Composition of pyroxenes in soils from (A) a highland site and (B) a mare site. Note the extrusion of the highland pyroxene field towards the low-Ca enstatite apex (after Apollo Soil Survey, 1974).

the compositional range of olivines is larger in highland soils than in mare soils.

Most soils contain some opaque minerals, mainly ilmenite. The ilmenite is probably largely basaltic in origin at both mare and highland sites. It is very variable in amount and can be quite abundant in some mare soils such as the Apollo 11 site where it forms a large percentage of the basement rock. A variety of other minerals such as spinels are present in the soils but in minor amounts only.

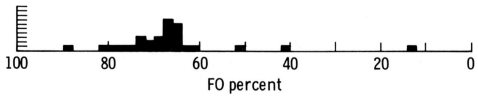

Fig. 6.19. Histogram of olivine compositions in an Apollo 14 soil, expressed as mole per cent forsterite (after Apollo Soil Survey, 1974).

Metallic Particles

Metal grains are present in very small numbers in all of the lunar soils. The particles are generally small, few being larger than 100 μm, and consist largely of kamacite and taenite and in some cases schreibersite and troilite. Some particles are free standing metal whereas others are associated with a silicate assemblage. Bulk chemistry of the particles indicates that they originate in three ways: (1) Many particles, particularly in mare areas, come from igneous rocks, (2) a proportion, however, are relatively unmodified meteoritic materials whereas (3) others are metallic spheroids which appear to be impact melted fragments of the meteoritic projectile (Goldstein *et al.*, 1972). The composition of the meteoritic materials suggests that they are chondritic in origin (Fig. 6.20).

Glass Particles

Glass particles are abundant in the lunar soils and provide some of the clearest insights into its provenance and to some extent its evolution. Compositionally and morphologically the glasses are extremely complex but they can conveniently be divided into two broad categories: (1) glasses which are essentially homogeneous and (2) agglutinates which are extremely inhomogeneous.

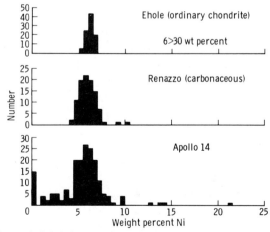

Fig. 6.20. A comparison of nickel distributions in metal fragments from Apollo 14 soils compared to metal separates from chondritic meteorites (after Goldstein *et al.*, 1972; *Proc. 3rd Lunar Sci. Conf., Suppl. 3, Vol. 1*, MIT/Pergamon).

Homogeneous Glasses

The homogeneous glasses are morphologically diverse. However, most are angular jagged fragments obviously derived from larger glass fragments by comminution. A smaller number of glass particles which caught the attention of many people following the first lunar mission have rotational forms. The particles range from perfect spheres, to oblate and prolate spheroids, to dumbbell (Fig. 21) and teardrop shapes (Fig. 6.22). Some particles have detrital rock or mineral fragments as cores (Fig. 6.22). Others have a small number of mineral fragments dispersed through otherwise homogeneous glass. Some glasses are vesicular with vesicles forming as much as 30 per cent of the particle volume. In the extreme some spheres are actually hollow bubbles which are as large as 2 cm in at least one case (Fig. 6.23).

Variations of the morphology of rotational glass forms may be explained in terms of the surface tension of the molten glass, and the angular velocity of the spinning glass mass as it is ejected from the impact crater. The developmental sequence, from a sphere controlled entirely by surface tension to a dumbbell form as angular velocity increases, is shown in Figure 6.24 (Fulchignoni *et al.*, 1971). Obviously, the survival of a rotational form

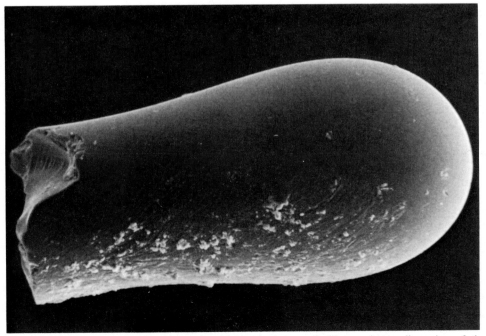

Fig. 6.21. A fractured rotational glass form from the lunar soil. The maximum dimension of the particle is 220 μm (NASA Photo S-72-52308).

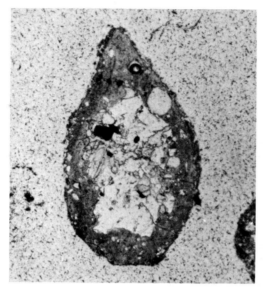

Fig. 6.22. A brown glass "teardrop" with a lithic core. The maximum dimension of the particle is 120 μm (NASA Photo S-72-46170).

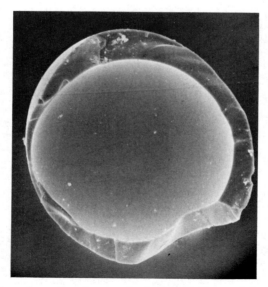

Fig. 6.23. Broken hollow glass sphere from an Apollo 12 soil. Note variation in wall thickness. The particle is approximately 100 μm is diameter.

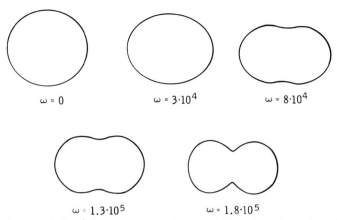

$\omega = 0$ $\omega = 3 \cdot 10^4$ $\omega = 8 \cdot 10^4$

$\omega = 1.3 \cdot 10^5$ $\omega = 1.8 \cdot 10^5$

Fig. 6.24. Evolution of the various rotational forms as a function of angular velocity (ω) (after Fulchignoni *et al.*, 1971; *Proc. 2nd Lunar Sci. Conf., Suppl. 2, Vol. 1*, MIT/Pergamon).

depends upon it having cooled below its transformational temperature before it intersects the lunar surface again. The available time will depend upon event magnitude and trajectory. Englehardt *et al.* (1973) has found evidence of cooling rates of 100°C s^{-1} for spherules from the Apollo 15 site, suggesting that relatively small impact events could produce rotational glass forms. From Figure 6.25 it can be seen that spherical and rotational glass forms occur throughout the soil but in a restricted grain-size range. The grain size restriction is probably connected with the cooling rate of the glass during transit and the surface tension of the glass. Larger glass forms simply do not cool fast enough to survive secondary impact with the lunar surface. Spheres and other rotational forms are not abundant in any size range and deserve considerably less attention than they received during the early phases of the Apollo program. However, it is worth comparing the distribution of spheres and rotational forms in the Apollo 16 soils with the Apollo 17 soils (Fig. 6.26). The Apollo 17 soils contain a large volcanic pyroclastic contribution most of which consists of glass spheres (see Chapter 4). Consequently concentrations of spheres and rotational forms of 5 per cent or more are not uncommon at the Apollo 17 site.

Homogeneous glasses come in a wide variety of colors — colorless, white, yellow, green, orange, red, brown and black. Most particles have darker colors such as brown or black. In general terms the darker glasses contain more Fe and Ti whereas the lighter glasses tend to be more aluminous. The refractive index of these glass particles ranges from 1.570 to 1.749 and like the color it varies directly with the total Fe and Ti content and inversely with the Al content (Chao *et al.*, 1970) (Fig. 6.27). The lighter colored glasses with a low refractive index tend to be more anorthositic or

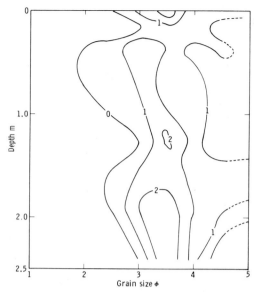

Fig. 6.25. Distribution of rotational glass forms (mainly spheres) as a function of both depth and grain size in the Apollo 16 deep drill core. Spheres tend to concentrate in a size range of from 3 to 4ϕ (125 to 62.5 μm)which suggests that scaling factors relating to surface tension and cooling rate control their formation. Concentrations are expressed as number per cent per 0.5ϕ grain size interval.

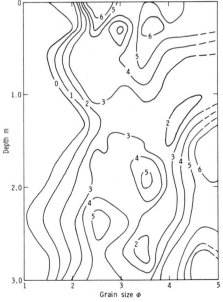

Fig. 6.26. Distribution of rotational glass forms as a function of depth and grain size in the Apollo 17 deep drill core. The higher concentrations of these particles in the Apollo 17 core by comparison with the Apollo 16 core is a concequence of the addition of large volumes of volcanic pyroclastic materials to the soil. Concentrations are expressed as number per cent per 0.5ϕ grain size interval.

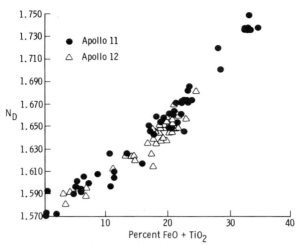

Fig. 6.27. Positive correlation of indices of refraction (N_D) of lunar glasses with the total FeO and TiO$_2$ content (after Chao, E., Boreman, J., Minkin, J., James, O. and Desborough, G., *Jour. Geophys. Res.*, 75: 7445-7479, 1970; copyrighted American Geophysical Union).

"highland" in composition, whereas the darker glasses with a high refractive index tend to be basaltic or "mare" in composition. The angular and rotational glass forms have the same range of colors and refractive indices, suggesting a common origin.

Chemistry of Homogeneous Glasses

The regular form of many homogeneous glasses suggests that they are impact melts sprayed onto the lunar surface during crater excavation, probably during the jetting phase. Terrestrial analogs indicate that in general the major element chemistry of the bedrock survives unchanged in an impact melt (Fredricksson *et al.*, 1974). Consequently, the homogeneous lunar soil glasses have been studied intensively in order to gain an insight into soil provenance and the composition of the lunar crust (Reid *et al.*, 1972a, 1972b; Reid, 1974; Ridley *et al.*, 1973; Apollo Soil Survey, 1971).

In Tables 6.8 and 6.9 the main compositional glass types are presented for two areas, the Apollo 15 site which is essentially a mare site, and the Apollo 14 site which is situated in the lunar highlands. The Apollo 15 data are particularly instructive in that the site is on the mare surface but a short distance from the Apennine Front. Both highland and mare rock compositions are, therefore, well represented. To add to the complexity of the Apollo 15 site the soils also contain a modest proportion of green-glass particles. These particles are basaltic in composition and are believed to be pyroclastic in origin.

A total of eleven compositional types occur among the homogeneous

TABLE 6.8

Average composition of homogeneous glass particles from the Apollo 15 soils (after Reid *et al.,* 1972b).

		MARE BASALT			
	Green Glass	Mare 1	Mare 2	Mare 3	Mare 4
SiO_2	45.43 (.63)	45.70 (2.01)	44.55 (1.94)	43.95 (1.89)	37.64 (2.27)
TiO_2	.42 (.06)	1.60 (.43)	3.79 (.98)	2.79 (.93)	12.04 (2.00)
Al_2O_3	7.72 (.33)	13.29 (1.96)	11.77 (1.88)	8.96 (1.20)	8.46 (1.07)
Cr_2O_3	.43 (.03)	.33 (.07)	.26 (.06)	.46 (.11)	.48 (.10)
FeO	19.61 (.84)	15.83 (2.16)	18.83 (2.14)	21.10 (1.46)	19.93 (1.76)
MgO	17.49 (.54)	11.72 (1.60)	8.84 (1.88)	12.30 (1.60)	10.49 (1.58)
CaO	8.34 (.41)	10.41 (.80)	10.46 (.94)	9.02 (.70)	8.81 (1.40)
Na_2O	.12 (.05)	.30 (.14)	.34 (.20)	.27 (.12)	.54 (.17)
K_2O	.01 (.02)	.10 (.07)	.13 (.11)	.05 (.03)	.13 (.05)
TOTAL	99.57	99.28	98.97	98.90	98.52
Percentage of total analyses	34.2	12.2	3.8	4.8	1.1

Numbers in parentheses are standard deviations.

glasses from the Apollo 15 site (Table 6.8) (Reid *et al.,* 1972), of which green glass particles are the most common. In terms of its normative composition it is a pyroxenite. Mg and Fe are high, Al, Ti and alkalis are low. The composition is essentially mare basalt with more Mg and Ti. The green glasses are mostly spheres and a few contain olivine needles (Fo_{74}). The origin of this glass has been discussed in Chapter 4. The general consensus at the present time is that they are pyroclastic materials formed in fire fountains during the flooding of the lunar mare. Orange glasses with a similar composition and morphology are present in the soils from the Apollo 17 site. At this site the pyroclastic contribution is 100 per cent of one soil sample, up to 22 per cent of other soil samples and results in anomalously high Zn concentrations (Rhodes *et al.,* 1974). Mixing model studies show considerable variation in the pyroclastic contribution in surface samples (Fig. 6.28). Spheres and rotational pyroclastic glass forms occur vertically through the soil blanket (Fig. 6.26).

The soils from the Apollo 14 site contain a similar range of compositional types, but the relative proportions of the glasses change in such a way that the amount of highland-derived Fra Mauro basalts increases whereas the mare basalts are of less importance.

A large number of mare glasses are present in the Apollo 14 and 15 soils and are characterized by a high Fe and low Al content. These glasses

Highland Basalt	FRA MAURO BASALT			'Granite' 1	'Granite' 2
	Low K	Moderate K	High K		
44.35 (2.38)	46.56 (1.78)	49.58 (1.43)	53.35 (2.82)	73.13 (1.36)	62.54
.43 (.37)	1.25 (.45)	1.43 (.30)	2.08 (.56)	.50 (.71)	1.18
27.96 (3.67)	18.83 (1.85)	17.60 (1.54)	15.57 (1.65)	12.37 (.87)	15.73
.08 (.04)	.20 (.05)	.17 (.03)	.12 (.06)	.35 (.49)	.03
5.05 (2.28)	9.67 (1.88)	9.52 (1.31)	10.25 (1.09)	3.49 (3.16)	6.67
6.86 (2.92)	11.04 (1.61)	8.94 (1.24)	5.77 (1.61)	.13 (.10)	2.51
15.64 (1.63)	11.60 (.93)	10.79 (.69)	9.57 (.77)	1.27 (.62)	6.86
.19 (.23)	.37 (.14)	.74 (.14)	1.01 (.42)	.64 (.16)	.98
.01 (.03)	.12 (.07)	.47 (.17)	1.11 (.40)	5.97 (.27)	3.20
100.92	99.66	99.24	98.83	97.85	99.70
6.6	15.0	16.5	5.3	.4	.2

resemble the composition of mare basalts and have been divided into four subtypes at the Apollo 15 site. This site is on the mare surface in Palus Putredinus and undoubtably many of the glasses are local in origin. However, Mare Imbrium, Mare Serenitatis and Mare Vaporum are nearby. The Mare 3 component is very similar to large basaltic rock samples returned from the

TABLE 6.9

Average composition of homogeneous glass particles from the Apollo 14 soils (after Apollo Soil Survey, 1974).

	Mare-type basaltic glass	Fra Mauro basaltic glass	Anorthositic gabbroic glass	Gabbroic anorthositic glass	'Granitic' glass	Low-silica glass
SiO_2	45.48 (2.64)	48.01 (1.75)	45.23 (1.00)	47.37 (2.52)	71.54 (7.60)	37.97
TiO_2	2.77 (2.44)	2.02 (.53)	.36 (.20)	.14 (.12)	.39 (.20)	.23
Al_2O_3	10.86 (3.10)	17.12 (1.83)	25.59 (1.08)	31.32 (1.34)	14.15 (4.21)	34.54
FeO	18.14 (2.93)	10.56 (1.37)	5.59 (1.00)	2.98 (1.51)	1.79 (2.16)	1.19
MgO	11.21 (4.25)	8.72 (1.82)	7.84 (1.49)	2.18 (1.38)	.70 (1.04)	5.57
CaO	9.56 (1.42)	10.77 (.86)	14.79 (.91)	14.78 (2.44)	1.97 (1.84)	20.39
Na_2O	.39 (.30)	.71 (.26)	.25 (.27)	.95 (.60)	.93 (.64)	0.00
K_2O	.32 (.42)	.55 (.28)	.12 (.25)	.22 (.22)	6.53 (2.44)	0.00
TOTAL	98.73	98.46	99.77	99.94	98.00	99.89

Numbers in parentheses are standard deviations.

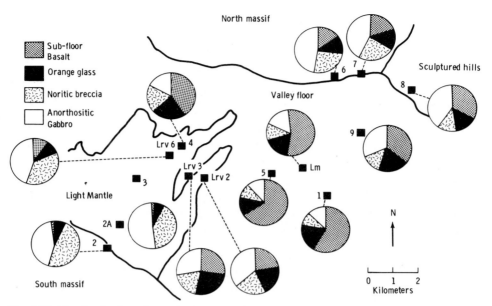

Fig. 6.28. The distribution of pyroclastic orange glass in the Apollo 17 surface soils compared to other local soil components (from Rhodes *et al.*, 1974, *Proc. 5th Lunar Sci. Conf., Suppl. 5, Vol. 2*, Pergamon).

site, and is probably of local origin. The other three types can not be directly attributed to particular sources, but Mare 1 glasses are similar to Mare Fecunditatis basalts whereas the Mare 4 glasses are similar to Mare Tranquillitatis basalts. The Mare 2 glasses are most similar to the Mare 3 glasses although there are significant differences such as higher Ca, Al and Sr and lower Fe and Mg content. It is possible that these glasses represent a more feldspathic mare basalt which was not sampled at the Apollo 15 site even though it may have been present.

The Fra Mauro basaltic glasses are a major constituent in the Apollo 15 soils and are comparable to those forming the most abundant glass type in the Apollo 14 soils. They are identical to brown ropy glasses called KREEP (for their higher potassium, rare earth element and phosphorus content) that were originally found in Apollo 12 soils (Meyer *et al.*, 1971) and are analagous to terrestrial tholeiitic basalts. One variant may be from the Apennine Front which is not a typical highland area. This is supported by orbital X-ray fluorescence data which indicates an Al/Si ratio similar to the low-K Fra Mauro basalt (Adler *et al.*, 1972a, 1972b). These glasses are probably representative of the pre-Imbrian crust.

Glasses characterized by a high Al_2O_3 content have been called highland basalt glasses (anorthositic gabbro). This is the anorthositic component generally held to be characteristic of the highlands. This component is low in the Apollo 15 soils which confirms arguments that the Apennine Front is not typical of the highlands.

Two glass types high in silica have the composition of a potash granite and a granodiorite. They have been found at other sites but are always in minor amounts. Their origin is unknown.

Provenance of Homogeneous Glasses

Reid *et al.* (1972b) have suggested that the pre-Imbrium surface consisted of Fra Mauro type rocks that had undergone some near surface differentiation. The excavation of the Imbrium Basin resulted in a redistribution of much of the differentiated sequence. The edges of Mare Imbrium were only flooded to a shallow depth by basalt, with the result that craters such as Copernicus, Aristillus and Autolycus penetrated the lava and excavated Fra Mauro material. The Apennine Front provided low-K material, perhaps from greater depths if we assume an inverted ejecta sequence. Apollo 14 sampled the same ejecta blanket but further from the crater rim and hence at shallower depth which accounts for the larger highland-basalt (anorthositic gabbro) content of the Apollo 14 soils. The anorthositic gabbro appears to be the major highland rock type.

The origin of the mare basalts is comparatively well understood and has been discussed in Chapters 1 and 4. Phase relations based on experimental work constrain the basaltic source to a depth of between 100 and 400 km (Ringwood and Essene, 1970; Green *et al.*, 1971). The highland rocks are, however, not so simply explained. It has been concluded from experimental studies that the highland rocks are not partial melts from the same source as the mare basalts (Walker *et al.*, 1973; Hodges and Kushiro, 1973), but that they probably evolved at shallower depths from feldspathic cumulates. The most comprehensive attempt to explain the relationship is that of Walker *et al.* (1973). They propose that the low-K and medium-K Fra Mauro basalts are derived by partial melting from more feldspathic rocks as outlined in Figure 6.29.

Agglutinates

Agglutinates have also been referred to as glazed aggregates (McKay *et al.*, 1971). or "glass-bonded aggregates" (Duke *et al.*, 1970) but agglutinates are much more complex than these terms suggests (McKay *et al.*, 1971, 1972; Lindsay, 1971, 1972, 1975). They are intimate mixtures of inhomogeneous dark-brown to black glass and mineral grains, many of which are partially vitrified. More than 50 per cent of most agglutinates is

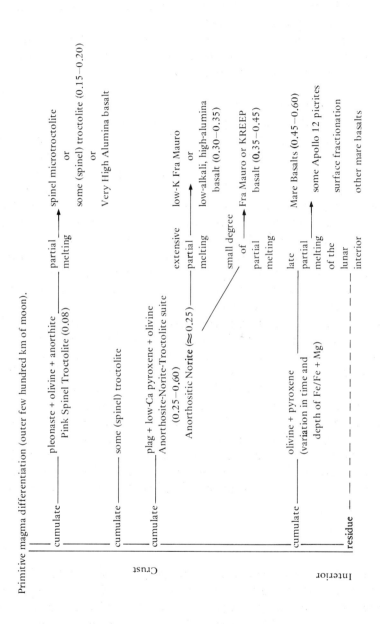

Fig. 6.29. Genetic relationships among highland rocks (from Walker *et al.*, 1973).

mineralic. Most are extremely complex in shape but have a general rounded form suggesting that their final shape was determined by the viscosity and surface tension of the fluid glass (Fig. 6.30). The surface of the grains has a coating of fine detrital fragments which gives them a dull saccaroidal texture (Figs. 6.31, 6.32). Some agglutinates are bowl shaped or take the form of rings or donuts (Fig. 6.31), suggesting that they are "pools" of melt formed in the bottom of microcraters in fine-grained lunar soil (McKay *et al.*, 1972; Lindsay, 1971, 1972c, 1975). Although some consist of dendritic glass projections radiating from a central mineral grain (Fig. 6.33) a few agglutinates are vesicular (particularly larger particles), most are massive and overall less vesicular than the homogeneous glasses.

Chemistry of Agglutinates

The chemical composition of individual agglutinates in a general way reflects the composition of the underlying bedrock. Thus agglutinates from the Apollo 12 site bear a general basaltic imprint (Table 6.10). Because of the inhomogeneity of these glasses more is to be gained from a study of their bulk chemistry than from the chemistry of individual particles. Table 6.11 shows the chemistry of the agglutinate and non-agglutinate fractions of four soils in order of increasing agglutinate content. There are significant differences in chemistry between the two fractions of the soils, and neither fraction is comparable chemically to a major homogeneous glass group (Rhodes *et al.*, 1975). The agglutinate fraction is not constant in composition but varies for a given element roughly in accordance with the bulk soil composition and the agglutinate content; this is particularly so in the case of trace elements. These variations are regular and systematic. That

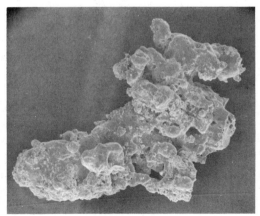

Fig. 6.30. Typical dust coated agglutinate. The maximum dimension of the particle is approximately 180 μm.

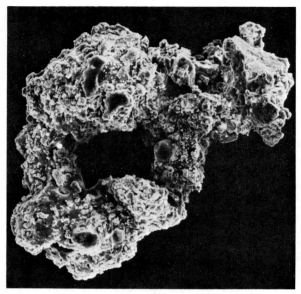

Fig. 6.31. Ring-shaped agglutinate. Such agglutinates support the concept that they are formed by micrometeoroid impact in the lunar soil. The maximum dimension of the particle is approximately 180 μm.

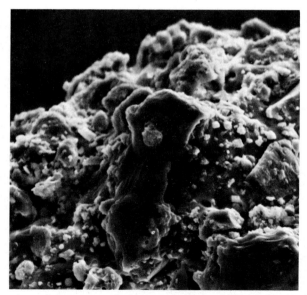

Fig. 6.32. Close-up view of the same agglutinate as in Figure 6.30 showing its dust coated surface Particles of plastically deformed glass adhere to the dust also.

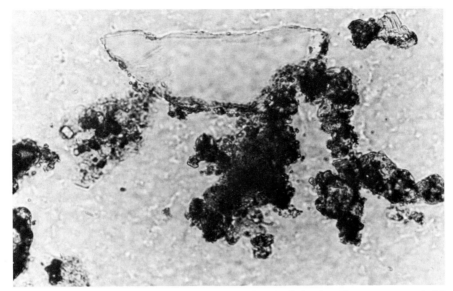

Fig. 6.33. An extremely delicate agglutinate consisting of a detrital mineralic core surrounded by radiating projections of dark-brown glass intimately mixed with fine grained mineralic materials. The maximum dimension of the particle is 230 μm (NASA Photo S-71-51084).

TABLE 6.10

Chemical analyses of three agglutinates from an Apollo 12 soil (after Chao *et al.*, 1970).

	1	2	3
SiO_2	46.7	42.2	44.8
TiO_2	2.31	3.34	2.60
Al_2O_3	15.4	9.8	15.6
FeO	12.4	18.3	13.7
MgO	7.9	12.0	7.8
CaO	10.7	8.9	11.6
Na_2O	0.56	0.32	0.54
K_2O	0.92	0.21	0.23
P_2O_5	0.62	0.19	0.17
MnO	0.10	0.12	0.12
Cr_2O_5	0.21	0.35	0.29
NiO	—	0.04	0.04
TOTAL	98	96	97

TABLE 6.11.

Major element chemistry of agglutinate and non-agglutinate fractions of four Apollo 16 soils (after Rhodes et al., 1975).

	65701		61241		61501		64421		
Sample sub sample	11	11	26	26	15	15	18	18	1
fraction	Agglu-tinate	Non-agglu-tinate	Agglu-tinate	Non-agglu-tinate	Agglu-tinate	Non-agglu-tinate	Agglu-tinate	Non-agglu-tinate	Bulk soil
wt. fraction	0.613	0.387	0.571	0.429	0.557	0.443	0.555	0.445	1.000
Major elements (wt. %)									
SiO_2	44.82	45.11	44.99	45.11	44.40	45.09	44.85	44.89	44.97
TiO_2	0.73	0.52	0.71	0.39	0.69	0.43	0.68	0.43	0.53
Al_2O_3	25.66	27.85	25.41	29.31	25.29	28.40	26.32	29.08	27.82
FeO	6.61	4.57	6.58	3.49	6.57	4.03	6.07	3.77	4.71
MnO	0.09	0.06	0.09	0.05	0.09	0.06	0.08	0.06	0.05
MgO	6.52	5.39	6.68	4.23	6.74	5.13	5.99	4.37	5.36
CaO	14.91	15.85	14.85	16.57	14.72	16.11	15.22	16.45	15.71
Na_2O	0.39	0.50	0.50	0.54	0.48	0.53	0.46	0.48	0.58
K_2O	0.15	0.12	0.12	0.09	0.13	0.10	0.12	0.09	0.11
P_2O_5	0.11	0.15	0.12	0.07	0.11	0.11	0.13	0.09	0.09
S	0.08	0.05	0.07	0.03	0.08	0.03	0.08	0.03	0.05
Total	100.07	100.17	100.12	99.88	99.30	100.02	100.00	99.74	99.98

is, there appears to be some form of chemical fractionation associated with agglutinate formation. In all cases the agglutinates are enriched in all of the ferromagnesian elements (Fe, Ti, Mg, Mn, Cr, Sc) and most of the lithophile elements (K, La, Ce, Sm, Yb, Lu, Hf, Th, Ta) and depleted in those elements which are characteristic of plagioclase (Al, Ca, Na, Eu). The agglutinates seem to be enriched in elements that igneous systems tend to concentrate either in mesotasis material or mafic minerals, especially clinopyroxene, but are excluded by plagioclase (Rhodes et al., 1975). Rhodes et al. (1975) have suggested that the chemical differences result from selective melting and assimilation of mesostasis material and mafic soil components, such as pyroxene and ilmenite, into the agglutinate glass by a multistage fluxing process. However, in light of the fact that the pyroxene/plagioclase ratio increases in finer fractions of the soil and the fact that agglutinates preferentially incorporate materials finer than 2.5ϕ (177 μm), it appears more likely that the chemical differences relate either to the different mechanical properties of the mineral components, or to selective sorting of the mineral components during transport and deposition.

It thus seems most likely that agglutinates are formed by shock melting and partial melting of fine-grained detrital materials by micrometeoroid impact (McKay *et al.*, 1972). Their morphology tends to suggest that an impact produces only one agglutinate. This observation has been used to obtain information about the ancient meteoroid flux (Lindsay and Srnka, 1975). As is discussed in more detail in a following section, an impacting meteoroid fuses about five times its own mass of target material with the result that the size distribution of agglutinates tends to be controlled by the mass frequency distribution of micrometeoroids. The longer a soil is exposed to the micrometeoroid flux the higher the content of agglutinates. The agglutinate content can be directly related to the exposure age of the soil as determined from fossil cosmic ray tracks (Fig. 6.34) (McKay *et al.*, 1972; Arrhenius *et al.*, 1971).

Grain Size of Agglutinates

   Agglutinates are constructional particles and contain an abundance of fine-grained detrital materials. The included particles are generally finer than 3ϕ (125 μm) and have a median size of close to 4.7ϕ (38 μm) (Lindsay, 1972c). The median grain size of included detritus varies from particle to particle depending upon the size of the agglutinate. The included particles are poorly sorted and have distributions similar to the fine tail of the bulk soil. An agglutinate thus removes finer detrital materials from circulation by incorporating them into a larger complex particle.

   Single agglutinates may be as large as 0ϕ (1 mm), although rarely, and comminuted agglutinate fragments occur in abundance in particle sizes as small as 6ϕ (16μm). However, most unbroken agglutinates occur in a narrow size range between about 2ϕ (250 μm) and 2.5ϕ (178 μm) (Fig. 6.35). As

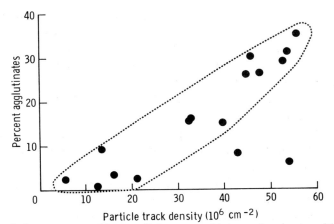

Fig 6.34. The agglutinate content of lunar soils increases with increasing density of fossil cosmic ray particle tracks, suggesting that exposure time at the lunar surface determines agglutinate content (adapted from McKay *et al.*, 1972; *Proc. 3rd Lunar Sci. Conf.*, *Suppl. 3, Vol. 1*, MIT/Pergamon).

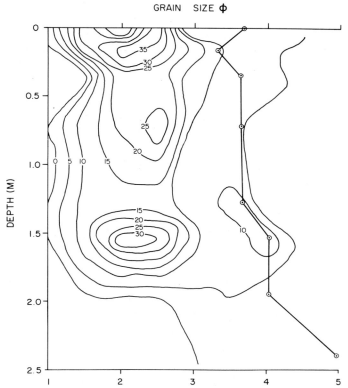

Fig. 6.35. Distribution of unbroken agglutinates as a function of grain size and depth in Apollo 16 drill core samples. Concentrations are expressed as number per cent per 0.5φ grain size interval. Most agglutinates are concentrated at a grain size of 2.5φ (177 μm) (from Lindsay, 1975).

discussed previously, the proportion of agglutinates in any one sample varies considerably apparently as a function of exposure age (McKay *et al.*, 1972). Agglutinate concentrations fluctuate, therefore, with depth (Fig. 6.35). Typically the size distribution of unbroken agglutinates has a mean of 2.45φ (184 μm) and is moderately to moderately-well sorted (δ_I = 0.80φ). The distribution is generally fine skewed (Sk_I = 0.189) the coarse end of the distribution apparently being truncated. The kurtosis of the distribution approaches log-normal at 0.924. Agglutination thus removes the fine end of the bulk soil grain-size distribution and shifts it to the coarse side of the mean grain size of the bulk soil.

The grain-size distribution of the agglutinates appears to be controlled by a complex of variables, but to a large extent it is a direct function of the meteoroid mass-frequency distribution. At the root-mean-square velocity of 20 km s[-1] a meteoroid fuses or vaporizes about 7.5 times its own mass of detrital materials (Gault *et al.*, 1972; Lindsay, 1975). Assuming that the detrital material enclosed in the agglutinate forms part of the solid-liquid-mixture transition, an agglutinate should be about 5 times the mass of the impacting meteoroid. The size distribution of agglutinates on the fine side of 2.2ϕ (217 μm) fits the model well. However, on the coarse side of 2.2ϕ the size distribution is depleted by comparison with the micrometeoroid flux. This implies that agglutination is inhibited by a second variable, which relates to scale (Lindsay, 1975). The most likely variables are viscosity and surface tension of the melt. Agglutinate formation relies on the ability of the melt to maintain its integrity during crater excavation. As event magnitude increases the probability that an agglutinate will form decreases, because the glass is dissipated as small droplets. Consequently the agglutinate grain-size distribution is fine skewed and by comparison with the meteoroid flux it is excessively deficient in coarser particles. The standard deviation of the agglutinate distribution cannot, therefore, be larger than the standard deviation of the meteoroid flux; it is in fact about half of the equivalent flux value.

The mean agglutinate size distribution suggests that most agglutinates are formed by micrometeoroids in the mass range 5.5×10^{-5} to 7.0×10^{-8} g. This mass range includes approximately 68 per cent of the mass and hence of the kinetic energy of the meteoroid flux. Obviously, agglutinate formation is one of the major processes active in the reworking of mature lunar soils.

The Genetic Environment of Agglutinates

Agglutinates are the product of extreme shock induced pressures of very short duration. Experimental data suggest that shock-produced melts are not formed in significant amounts in the lunar soil at pressures below 100 kbar (Gibbons *et al.*, 1975). At 250 kbar only 1 to 2 per cent of melt is formed, and some of the plagioclase is transformed to maskelynite. Plagioclase is completely transformed to maskelynite at 288 kbar and about 5 per cent of the mass is shock-vitrified. The melt content increases to approximately 20 per cent at 381 kbar. When pressures are elevated to 514 kbar there is approximately 50 per cent glass and virtually all plagioclase is melted. The intergranular melt is vesicular and there are prominent schlieren (Gibbons *et al.*, 1975). The experimental work suggests that most natural agglutinates are probably formed at shock induced pressures of between 381 kbar and 514 kbar.

Mixing Models and End Members

It is readily apparent from the homogeneous glass compositions that the lunar soil is a complex mixture of several major lithologies (Chao *et al.*, 1970; Apollo Soil Survey, 1971; Reid *et al.*, 1972, 1973; Lindsay, 1971). However, by themselves the glasses do not give us more than a general feeling for the relative importance of the various components or end members, nor do they offer complete information concerning variations in the mixture at a particular locality. This problem can be approached directly through mixing models. A general mixing model for N end members can be defined as:

$$O_i = \sum_{j=1}^{N} a_j X_{ij} + e_i \qquad (6.3)$$

where O_i is the observed percentage of the ith oxide (or element) in the bulk soil, a_j is the proportion of the jth end member contained in the soil, X_{ij} is the percentage of the ith oxide (or element) in the jth end member and e_i is the random error term associated with the ith oxide or element. End-member proportions can thus be estimated by a least-squares solution and the fit of the model can be assessed from the percentage of total corrected sums of squares accounted for.

Several attempts have been made to estimate end-member proportions using models similar to equation 6.3. The basic problem in this approach is the determination of meaningful end member compositions. Some authors have proposed as few as two end members (Hubbard *et al.*, 1971), a mare basalt and a Fra Mauro basalt (KREEP) end member; others have proposed as many as five or six end members (Goles *et al.*, 1971; Meyer *et al.*, 1971; Schonfeld and Meyer, 1972). These end members have been proposed largely on intuitive reasoning and some have little statistical significance. Lindsay (1971) attempted in essence to reverse the mixing model calculations by using factor analysis to separate statistically meaningful end-member groups from chemical analyses of glass. This led to the recognition of three end members (1) an average mare basalt, (2) an average anorthosite, (3) an average Fra Mauro basalt (KREEP). These end members agree well with the Reid *et al.* (1972b) model for the provenance of homogeneous glasses. One end member represents the mare basalt composition, one the typical anorthositic crustal material and the third the deeper materials excavated by the Imbrian event. The results of mixing-model calculations in Fig. 6 show that the Apollo 12 mare soils are dominated by the mare basalt and member, but all have varying proportions of the other two end members. Hubbard *et al.* (1971) believed the Fra Mauro basalt (KREEP) component of the Apollo 12 soils to be ejecta from Copernicus. The fact that soil 12033 in Fig. 6.36 consists entirely of local mare basalt material

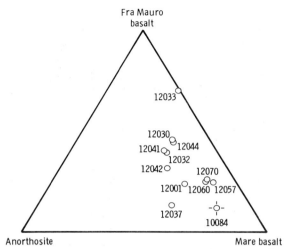

Fig. 6.36. Ternary plot of mixing models for Apollo 11 and 12 soil samples (after Lindsay, 1971).

and Fra Mauro basalt tends to support this suggestion, and emphasize the complex crustal model proposed by Reid *et al.* (1972b), which suggests that the Fra Mauro component was excavated from beneath a thin mare basalt layer and distributed widely on the lunar surface.

Texture of the Lunar Soil

Grain Size of Lunar Soils

Typically the lunar soils are fine grained and poorly sorted. They are moderately coarse skewed and near log normal in terms of kurtosis. There is, however, a considerable variability in the grain size parameters of samples from any one site (Duke *et al.*, 1970; Lindsay, 1971, 1972c, 1973, 1974, 1975; McKay *et al.*, 1972, 1974; Heiken *et al.*, 1973; Butler and King, 1974; Butler *et al.*, 1973; King *et al.*, 1971, 1972).

The mean grain size of the soils ranges from a coarse extreme of -1.4ϕ (380 μm) for a single sample at the Apollo 12 site to an extremely fine grained sample with a mean of 4.969ϕ (32 μm) at the Apollo 16 site. The former is believed to be primary detrital material excavated from the bedrock beneath the soil whereas the latter appears to be the product of gaseous sorting, perhaps in a base surge. However, most soils have means close to 4ϕ (Fig. 6.37). For example, the grand mean grain size of 50 samples from the Apollo 15 site is 4.019 $\pm$ 0.334ϕ (62 μm) (Lindsay, 1973). Variations in mean grain-size are not random but form part of a time

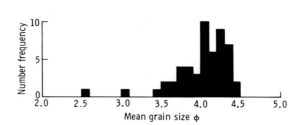

Fig. 6.37. Frequency distribution of the graphic-mean grain size of Apollo 15 core samples (after Lindsay, 1973; *Proc. 4th Lunar Sci. Conf., Suppl. 4, Vol. 1,* Pergamon).

sequence relating to the amount of reworking each soil layer has undergone. In Figure 6.38 the soil parameters are plotted in stratigraphic context and it can be seen that the mean grain size of the soil decreases upwards in a regular manner, with minor erratic excursions which are probably due to the introduction of either older coarse soils or freshly excavated bedrock material (Lindsay, 1973). Further, the mean grain size of the soil is strongly related to the content of agglutinates on the coarse side of the mean (Fig. 6.39) (McKay *et al.,* 1974). As the soil blanket evolves the accumulation rate decreases and the soil is subjected to longer periods of reworking by micrometeoroids resulting in an increased agglutinate content and a finer mean grain size. It is also apparent from Figure 6.39 that the scatter about the regression line increases as the grain size of the soil decreases. This scatter probably relates to random destruction of large numbers of agglutinates by larger layer forming impact events. More data concerning this process is available from studies of skewness.

The standard deviation or sorting of the lunar soils likewise varies but tends toward values close to 1.8ϕ. The most poorly sorted soils are mixed soils from around Cone Crater at the Apollo 14 site, which have standard deviations in excess of 4ϕ. The most fine-grained soil from the Apollo 16 site is also the most well sorted with a standard deviation of 0.990ϕ. Like the mean, the standard deviation conforms to a time sequence in which the soil becomes better sorted with time (Fig. 6.38) (Lindsay, 1973). Further there is a strong linear relationship between mean and standard deviation which shows that finer soils are better sorted (Fig. 6.40) (Lindsay, 1971, 1972, 1973, 1974; McKay *et al.,* 1974; Heiken *et al.,* 1973). Similarly the better sorted soils contain a greater abundance of agglutinates on the coarse side of the mean.

To this point the relationships are relatively straight forward; with increasing time soils become finer grained, better sorted and show evidence of increased reworking in the form of agglutinates. However, when we attempt to place skewness in the same context the picture becomes more complex. Skewness does not relate in a simple linear way to mean and

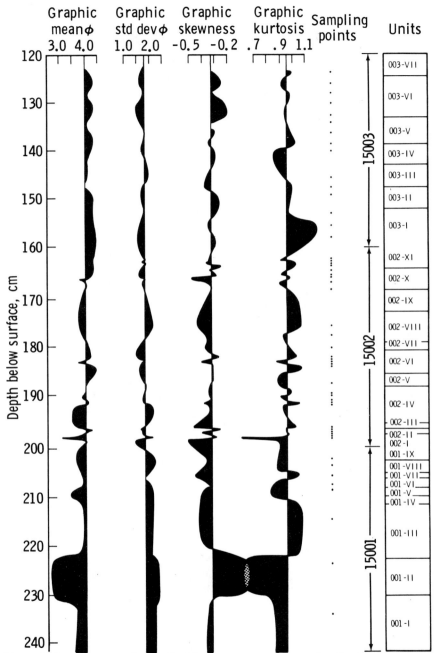

Fig. 6.38. Grain-size parameters for Apollo 15 deep-core samples as a function of depth below the lunar surface. Black areas indicate deviations about mean values (after Lindsay, 1973; *Proc. 4th Lunar Sci. Conf., Suppl. 4, Vol. 1*, Pergamon).

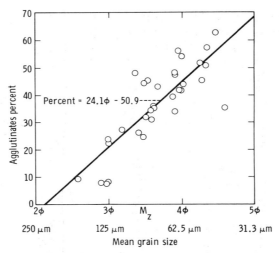

Fig. 6.39. Graphic mean grain size as a function of agglutinate content in the 3.5ϕ to 2.7ϕ (90-150 μm) fraction of the lunar soil (from McKay *et al.*, 1974; *Proc. 5th Lunar Sci. Conf., Suppl. 5, Vol. 1,* Pergamon).

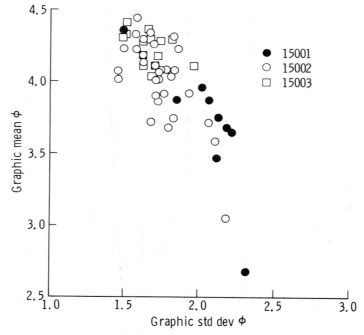

Fig. 6.40. Mean grain size as a function of standard deviation for soils from the Apollo 15 deep drill core (data from Lindsay, 1973).

standard deviation but is controlled by a double function (Fig. 6.41). The formation of agglutinates on the coarse side of the mean could be expected to produce soils that gradually become more negatively (coarse) skewed as the mean and standard deviation become smaller. This relationship does exist but is relatively weak (Fig. 6.38). The second function suggests that a second set of variables is operating to determine the skewness of many soils. The process involved has been called "cycling" (Lindsay, 1975).

Agglutinates are extremely delicate, fragile particles that can be destroyed by comminution much more readily than rock or mineral fragments of equivalent size. The relationship between agglutinate content and exposure age indicates that under normal conditions micrometeoroid reworking produces more agglutinates than it destroys by comminution (Lindsay, 1972, 1975; McKay *et al.*, 1974). This implies that the second function controlling skewness must be due to the destruction of large numbers of agglutinates possibly by larger layer-forming impact events (Lindsay, 1975). The destruction of agglutinates reduces the mean grain-size of the soil and makes the grain-size distribution more symmetrical (i.e. less negatively skewed). The larger the agglutinate content of the soil the more dramatic the effect. Once excavated to the surface micrometeoroid reworking again takes effect and the grain size parameters begin to converge on the ideal evolutionary path. Cycling is thus a series of random excursions during which agglutinates are first formed at the expense of the fine tail of the grain size distribution by micrometeoroid reworking, and are then instantaneously crushed and shifted back to the fine tail again.

Kurtosis appears unrelated to all other grain size parameters and there is no stratigraphic trend to suggest any evolutionary sequence as is seen in the mean and standard deviation (Fig. 6.38). For a log-normal grain-size distribution the kurtosis is 1.0 and most lunar soils fall close to this value. Some soils, however, have kurtosis values of less than 1.0 indicating that the size distribution is excessively broad compared to a log-normal curve. These soils appear to be mixtures in which two or more soils with different grain-size parameters have been brought together to form a broad bimodal or polymodal distribution. Thus, in Figure 6.38 the random deviations in the kurtosis values indicate the introduction of either more primitive soils from deeper in the soil blanket, or freshly excavated bedrock materials (Lindsay, 1975).

Shape of Soil Particles

The shape of soil particles varies considerably, from the smooth spherical form of glass droplets to the extremely irregular and complex

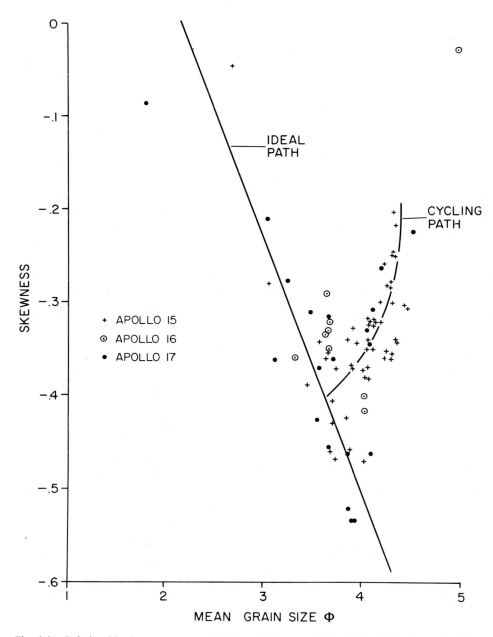

Fig. 6.41. Relationships between mean grain-size and skewness. This relationship is determined by two functions, one an ideal evolutionary path resulting from the gradual addition of agglutinates, the other due to cycling as agglutinates are alternately crushed by larger events and then restored by micrometeoroid reworking at the lunar surface, bringing the skewness back to the ideal path (from Lindsay, 1975).

shapes of agglutinates. Between these two extremes are the blocky angular comminuted rock, mineral and glass fragments. The overall mean sphericity of the soil particles is 0.78 (a value of 1.0 being a perfect sphere). However, perhaps predictably the sphericity of particles varies markedly with grain size and depth in the lunar soil (Lindsay, 1972c, 1974, 1975).

If we look at the sphericity of detrital particles forming a primitive soil (consisting largely of comminuted rock and mineral fragments) we find that sphericity changes in a regular linear manner with grain size. Smaller particles become increasingly more spherical. If we then investigate a texturally mature soil we find that a sinusoidal distribution is superimposed on the general linear trend. Two zones of depressed sphericity develop; one at between 0ϕ (1 mm) and 3ϕ (125 μm), the other at between 5ϕ (31 μm) and 7ϕ (8 μm). The midpoint of these two zones lies close to one standard deviation either side of the mean grain-size of the bulk soil. This relationship is seen to best advantage in Figure 6.42 where sphericity is shown as a function of both grain size and depth in the soil. The zones of reduced sphericity coincide with whole agglutinates on the one hand and comminuted agglutinates on the other. The relationship between particle shape and size thus relates entirely to the textural evolution of the soil and in particular to micrometeoroid reworking at the lunar surface.

Textural Evolution of the Lunar Soil

The texture of the lunar soil evolves in a regular manner in a direct response to continued reworking by the meteoroid flux (Lindsay, 1971, 1972c, 1973, 1975; McKay *et al.,* 1974; Heiken *et al.,* 1973; Butler and King, 1974). The two main dynamic processes responsible for this developmental sequence are (1) comminution and (2) agglutination. The textural maturity of the lunar soil is determined almost entirely by the balance between these two opposing processes, one destructive the other constructive. As a result of this complex interaction between the soil and the meteoroid flux the soil passes through three transitional evolutionary stages (Lindsay, 1975). For convenience the stages are designated (1) the comminution dominated stage (2) the agglutination dominated stage and (3) the steady state or cycling stage (Fig. 6.43).

The Comminution Dominated Stage

Because of the extreme age of the lunar surface few soils have survived that could be assigned to this evolutionary stage. Energetic impact events striking lunar bedrock produce coarse grained and poorly sorted ejecta. When such loose particulate materials are subjected to repetitive fracture the grain size distribution is gradually modified in such a way that it asymptotically approaches a log-normal form (Kolmogoroff, 1941; Halmos, 1944; Epstein, 1947). The mean grain-size of the particulate materials is

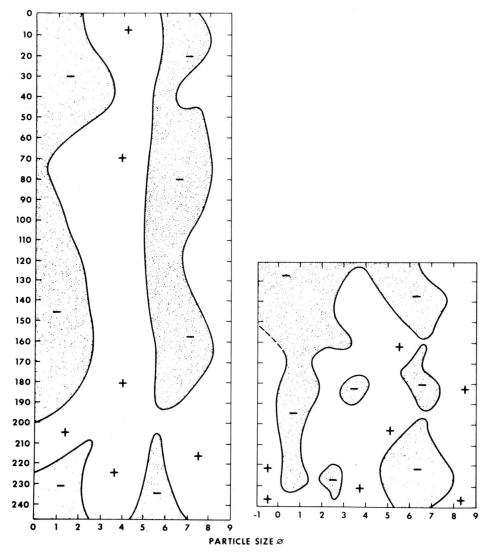

PARTICLE SIZE $\varnothing$

Fig 6.42. Sphericity of detrital particles as a function of grain size and depth in the lunar soil at the Apollo 16 site (left) and the Apollo 15 site (right). Sphericity is expressed as a deviation about the overall mean value of 0.78. Stippled areas are zones of reduced sphericity (from Lindsay, 1975).

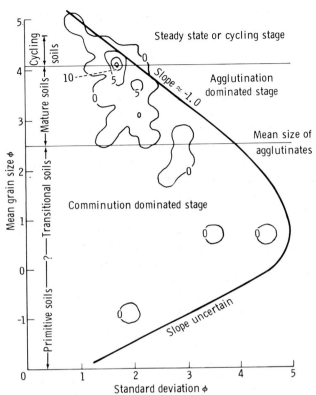

Fig. 6.43. Stages in the evolution of lunar soil as shown by change in mean and sorting of grain-size data. The solid line indicates the evolutionary path for an idealized soil. Contours show density in per cent of soil samples per 0.5ϕ square. Note that most soils are finer than the mean agglutinate size and they concentrate at a mean of 4.1ϕ, the stable point between the two glass modes (after Lindsay, 1974, 1975).

gradually reduced and they become more poorly sorted. These primitive soils are coarse grained with means in excess of 0ϕ (1 mm). During this stage of soil development comminution is the only effective process as most meteoroids interact with single detrital particles. The particles of soil are so large that most micrometeoroid impacts are either involved in catastrophic rupture of grains or they simply form pits on the larger grains.

Experimental studies suggest that cratering occurs in loose particulate material only when the particle mass is less than an order of magnitude larger than the impacting projectile. Thus comminution will dominate soil dynamics until the soil particles are reduced to masses of the order of 10^{-5} g, that is an order of magnitude larger than the mean mass of the meteoroid flux (10^{-6} g). This suggests that the soils would have a mean grain-size of the order of 3.4ϕ (93 μm). At the very least it is unlikely that agglutination can become effective until the mean grain-size of the soil is less than 2.5ϕ, the mean grain-size of the agglutinate distribution.

The Agglutination Dominated Stage

Most of the soils presently at the lunar surface appear to be passing through this stage of development. Once the mean grain-size of the soil is small enough for agglutination to be effective, it begins to balance the effects of communition (Duke, 1970; Lindsay, 1971, 1972, 1974, 1975; McKay *et al.*, 1974). This results in a continuing decrease in the mean grain-size as larger lithic and mineral grains are comminuted; the standard deviation of the grain size distribution is reduced causing the ideal evolutionary path in Figure 6.43 to reverse its slope. During this stage there is a strong correlation between mean grain size and standard deviation, and skewness tends to become more negative (Lindsay, 1971, 1972, 1975; McKay *et al.*, 1974; Heiken *et al.*, 1973). The glass particles produced by the two opposing processes (agglutination and comminution) are formed in two narrow size-ranges. Agglutinates concentrate at approximately 2.5ϕ while comminuted agglutinates accumulate at approximately 6ϕ. The distribution of these irregularly shaped particles is clearly shown in Figure 6.42. The mean grain-size of most lunar soils lies between these two glass modes (Lindsay, 1972). Ultimately, as the coarse rock and mineral fragments are broken down by comminution, the mean grain size is forced to lie between the two glass modes at close to 4ϕ and the standard deviation of the soils is restricted to about 1.7ϕ, or roughly coincident with the glass modes. At this point the soils are approaching a steady state. However, agglutination still dominates in that more agglutinates are formed than are destroyed by comminution.

The Steady State Stage

As the evolving soil moves closer to the point of stabilization, where the mean lies between the glass modes, the grain size parameters become more and more dependent on the presence of the agglutinate mode.

At this point a soil may enter a cycling mode. Agglutinates are fragile and readily crushed so that when a soil is redistributed by a larger layer-forming impact event a large proportion of the agglutinate population is destroyed. Removal of the agglutinates causes the mean and standard deviation to decrease and the skewness to become less negative. That is, the soil tends to take on the grain-size parameters of the second glass-mode. The soil is then exposed at the lunar surface to the micrometeoroid flux and agglutination once again become effective, pulling the grain size parameters back to the steady state ideal. Soils at this stage have been called steady state soils (Lindsay, 1975) or equilibrium soils (McKay *et al.*, 1974).

This model represents an ideal situation where processes are assumed to be operating in a closed system. In reality the soils are continually reworked and soils from two or more stages are continually being mixed. Soils from earlier erosional episodes may be fossilized by burial and later reexcavated to mix with more mature soils at the lunar surface (Lindsay, 1974, 1975; McKay *et al.*, 1975). At the same time larger impact events may penetrate the soil blanket and excavate bedrock material. The result is a complex interbedded sequence of mixed soils which gradually converge to a steady state upward through the stratigraphy (Fig. 6.44). The effects of mixing are well illustrated by the random deviations of kurtosis in Figure 6.38.

Soil Maturity

The term "maturity", when applied to detrital material, is generally used in the sense of a measure of progress along some predetermined path. If we consider sand accumulating on a beach we see several processes operative in modifying the texture and composition of the detrital materials. Wave action abrades, rounds and sorts the detrital particles while less stable minerals deteriorate in the aqueous environment and are removed. The end product is a well-sorted sand consisting of well-rounded quartz grains. In the case of beach sand the proportions of detrital quartz can be used as a measure of maturity. Thus in Pettijohns (1957) concept of maturity monomineralic quartz sand is "the ultimate end product" to which beach

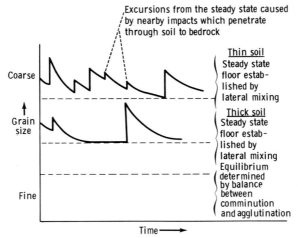

Fig. 6.44. Diagramatic representation of the relationship between grain size, soil thickness and time (from McKay *et al.*, 1974; *Proc. 5th Lunar Sci. Conf., Suppl. 5, Vol. 1*, Pergamon).

sand is "driven by the formative process that operate upon it."

To apply Pettijohns (1957) concept to the waterless depositional environment of the lunar soil requires a deeper insight into what we are actually attempting to measure with a maturity index. The answer is simply the total energy expended in the formation of the sediment (Lindsay 1971, 1972c, 1974). The formitive processes operating on the lunar soil in large part relate to the kinetic energy of the meteoroid flux at the lunar surface. Each impact on the lunar surface releases a significant proportion of its kinetic energy in the form of heat energy (Gault and Heitowit, 1963; Braslau, 1970). In turn a proportion of this heat energy is expended in fusing some of the detrital materials excavated by the impact (see Chapter 3). Since the soil is continually reworked by the meteoroid flux the glass content of the soil increases with time (Lindsay, 1971, 1972c, 1975). Thus the amount of glass in some soils could be used as an indicator of soil maturity.

In this light two indices of soil maturity have been proposed. Lindsay (1971) proposed that the total glass-content of the soil is a measure of mineralogic maturity within the confines of Pettijohns (1957) definition. The total glass-content of the soil must increase with time. However, the rate of increase in the glass content will decline exponentially with time as the accumulation rate decreases and more energy is expended in remelting pre-existing glass. The concept was tested using the known radiometric ages of the substrate at different sites, and it was found that soils overlying older substrates contained more glass. When plotted against grain size, however, it was found that soil texture was apparently unrelated to mineralogic maturity suggesting that textural maturity evolves in a more complex manner. In view of the steady-state or cycling stage of soil evolution this is understandable.

In a subsequent definition of maturity McKay et al. (1972) have suggested the use of agglutinates as a maturity index. They found that the agglutinate content of the soil correlates with its exposure age or residence time on the lunar surface. However, the agglutinate content of the soil also correlates with the grain-size parameters, suggesting that it is a measure of textural maturity rather than a measure of total energy (Lindsay, 1974; McKay et al., 1974). In the light of the evolutionary model discussed in the preceding section, it can be seen that agglutinates can only be used as an effective measure of maturity during the agglutination dominated stage of soil devolopment. In the comminution dominated stage agglutinates are not formed whereas in the cycling stage agglutinates are destroyed as fast as they are formed in random excursions away from the "ideal" mature soil. Some of these problems may be overcome if the total agglutinate content of the soil could be evaluated in the way in which Rhodes et al. (1975) have evaluated the agglutinate chemistry.

In summary, the total glass content of the soil must continually increase with time and offer a measure of total energy involved in the development of the soil. The agglutinates offer a more sensitive index of textural maturity during a period of soil development. However, very young or very old soils will not conform to the agglutinate model.

Energy Partitioning and the Flux of Detrital Materials

The energy released by hypervelocity impact at the lunar surface is partitioned in a complex way. First, and possibly most obviously, the kinetic energy of an individual impact event is released through heating and either fusing or vaporizing target and projectile, comminuting the substrate and finally ejecting the heated and comminuted materials. The kinetic energy of the flux is also partitioned in different ways according to the mass spectrum. Different portions of the mass spectrum contribute to or modify the soils in different ways (Fig. 6.2). It is as yet difficult to evaluate these variables but some order of magnitude estimates have been made (Lindsay, 1975).

The post-mare meteoroid flux at the lunar surface produces a primary sediment flux of the order of 2.78×10^{-7} g cm^{-2} yr^{-1} which represents an erosional efficiency of approximately 1.6 per cent (Gault *et al.*, 1972; Lindsay, 1975). The sediment flux per unit area entering the earth's oceans is 175 times larger than the lunar sediment flux.

Bedrock material appears to be excavated most efficiently by meteoroids larger than 10^3 to 10^4 g. In contrast, agglutination is caused by micrometeoroids in the mass range of 10^{-7} to 10^{-4} g or by about 68 per cent of the flux mass. Layer forming events appear to be the product of meteoroids larger than 7 g or less than 1 per cent of the flux mass.

Despite the tenuous nature of the energy source and the large inefficiencies involved, the meteoroid flux has, over several aeons, produced an extensive dynamic sedimentary body over most of the lunar surface. This sedimentary body probably contains stratigraphic information covering most of the life time of the solar system and perhaps a record of events of galactic magnitude.

References

Adler, I., Trombka, J., Gerard, J., Lowman, P., Schmadebeck, R., Blodgeth, H., Eller, E., Yin, L., Lamothe, R., Gorenstein, P. and Bjorkholm, P., 1972a. Apollo 15 geochemical X-ray fluorescence experiments. Preliminary report. *Science, 175*: 436-440.

Adler, I., Trombka, J., Gerard, J., Lowman, P., Schmadebeck, R., Blodgeth, H., Eller, E., Yin, L., Lamothe, R., Gorenstein, P., Bjorkholm, P., Gursky, H., Harris, B., Golub, L. and Harnden, R. F., 1972b. X-ray fluorescence experiment. In: *Apollo 16 Preliminary Science Report*, NASA SP- 315, 19-1 to 19-14.

Apollo Soil Survey, 1971. Apollo 14: nature and origin of rock types in soil from the Fra Mauro Formation. *Earth Planet. Sci. Lett., 12*: 49-54.

Arrhenius, G., Liang, S., Macdougall, D., Wilkening, L., Bhandari, N., Blat, S., Lal, D., Rajagopalan, G., Tamhane, A. S. and Benkatavaradan, V. S., 1971. The exposure history of the Apollo 12 regolith. *Proc. Second Lunar Sci. Conf., Suppl. 2, Geochim. Cosmochim. Acta, 3*: 2583-2598.

Apollo Soil Survey, 1974. Phase chemistry of Apollo 14 soil sample 14259. *Modern Geol., 5*: 1-13.

Braslau, D., 1970. Partitioning of energy in hypervelocity impact against loose sand targets. *Jour. Geophys. Res., 75*: 3987-3999.

Butler, J. C., Greene, G. M. and King, E. A., 1973. Grain size frequency distributions and modal analyses of Apollo 16 fines. *Proc. Fourth Lunar Sci. Conf., Suppl. 4, Geochim. Cosmochim. Acta, 1*: 267-278.

Butler, J. C. and King, E. A., 1974. Analysis of the grain size-frequency distributions of lunar fines. *Proc. Fifth Lunar Sci. Conf., Suppl. 5, Geochim. Cosmochim. Acta, 3*: 829-841.

Carrier, W. D., Mitchell, J. K., and Mahmood, A., 1974. Lunar soil density and porosity. *Proc. Fifth Lunar Sci. Conf., Suppl. 5, Geochim. Cosmochim. Acta, 3*: 2361-2364.

Chao, E. T. C., Boreman, J. A., Minkin, J. A., James, O. B. and Desborough, G. A., 1970. Lunar glasses of impact origin: physical and chemical characteristics and geologic implications. *Jour. Geophys. Res., 75*: 7445-7479.

Cherkasov, I. I., Vakhnin, V. M., Kemurjian, A. L., Mikhailov, L. N., Mikheyev, V. V., Musatov, A. A., Smorodinov, M. J. and Shvarey, V. V., 1968. Determination of physical and mechanical properties of the lunar surface layer by means of Luna 13 automatic station: *Tenth Plenary Meeting*, COSPAR, London.

Christensen, E. M., Batterson, S. A., Benson, H. E., Choate, R., Jaffe, L. D., Jones, R. H., Ko, H. Y., Spencer, R. L., Sperling, F. B. and Sutton, G. H., 1967. Lunar surface mechanical properties: Surveyor III Mission Report, Part II, Scientific Results: *California Inst. Technology Jet Propulsion Lab. Tech. Rept. 82-1177*, 111-153.

Cooper, M. R., Kovach, R. L., and Watkins, J. S., 1974. Lunar near-surface structure. *Rev. Geophys. Space Phys., 12*: 291-308.

Duke, M. B., Woo, C. C., Sellers, G. A., Bird, M. L. and Finkelman, R. B., 1970. Genesis of soil at Tranquillity Base. *Proc. Apollo 11 Lunar Sci. Conf., Suppl. 1, Geochim. Cosmochim. Acta, 1*: 347-361.

Duke, M. B. and Nagel, J. S., 1975. Stratification in the lunar regolith — a preliminary view. *The Moon, 13*: 143-158.

Englehardt, W. von, Arndt, J. and Schneider, H., 1973. Apollo 15: evolution of the regolith and origin of glasses. *Proc. Fourth Lunar Sci. Conf., Suppl. 4, Geochim. Cosmochim. Acta, 1*: 239-249.

Epstein, B., 1947. The mathematical description of certain breakage mechanisms leading to the logarithmic-normal distribution. *Jour. Franklin Inst., 44*: 471-477.

Finkelman, R. B., 1973. Analysis of the ultrafine fraction of the Apollo 14 regolith. *Proc. Fourth Lunar Sci. Conf., Suppl. 4, Geochim. Cosmochim. Acta, 1*: 179-189.

Fredricksson, K., Brenner, P., Nelen, J., Noonan, A., Dube, A. and Reid, A., 1974. Comparitive studies of impact glasses and breccias. *Lunar Science—V*, The Lunar Science Inst., Houston, Texas, 245-247.

Friedman, G. M., 1958. Determination of sieve-size distribution from thin-section data for sedimentary petrological studies. *Jour. Geology, 66*: 394-416.

Friedman, G. M., 1962. Comparison of moment measures for sieving and thin-section data in sedimentary petrographic studies. *Jour. Sediment. Pet., 32:* 15-25.

Fulchignoni, M., Funiciello, R., Taddeucci, A. and Trigila, R., 1971. Glassy spheroids in lunar fines from Apollo 12 samples 12070,37; 12001,73; and 12057,60. *Proc. Second Lunar Sci. Conf., Suppl. 2, Geochim. Cosmochim. Acta, 1:* 937-948.

Heiken, G. H., McKay, D. S. and Fruland, R. M., 1973. Apollo 16 soils: grain size analyses and petrography. *Proc. Fourth Lunar Sci. Conf., Suppl. 4, Geochim. Cosmochim. Acta, 1:* 251-265.

Gault, D. E., Quaide, W. L., Oberbeck, V. R., and Moore, H. J., 1966. Lunar 9 photographs: evidence for a fragmental surface layer. *Science. 153:* 985-988.

Gault, D. E. and Heitowit, E. D., 1963. The partitioning of energy for hypervelocity impact craters formed in rock. *Proc. Sixth Hypervelocity Impact Symp.,* Cleveland, Ohio, *2:* 419-456.

Gault, D. E., Horz, F., and Hartung, J. B., 1972. Effects of microcratering on the lunar surface. *Proc. Third Lunar Sci. Conf., Suppl. 3, Geochim. Cosmochim. Acta, 3:* 2713-2734.

Gault. D. E., Horz, F., Brownlee, D. E. and Hartung, J. B., 1974. Mixing of the lunar regolith. *Proc. Fifth Lunar Sci. Conf., Suppl. 5, Geochim. Cosmochim. Acta, 3:* 2365-2386.

Gibbons, R. V., Morris, R. V., Horz, F. and Thompson, T. D., 1975. Petrographic and ferromagnetic resonance studies of experimentally shocked regolith analogues. *Proc. Sixth Lunar Sci. Conf., Geochim. Cosmochim. Acta,* (in press).

Goldstein, J. I., Axon, H. J. and Yen, C. F., 1972. Metallic particles in Apollo 14 lunar soils. *Proc. Third Lunar Sci. Conf., Suppl. 3, Geochim. Cosmochim. Acta, 1:* 1037-1064.

Goles, G. G., Duncan, A. R., Lindstrom, D. J., Martin, M. R., Beyer, R. L., Osawa, M., Randle, K., Meek, L. T., Steinborn, T. L. and McKay, S. M., 1971. Analyses of Apollo 12 specimens: compositional variations, differentiation processes, and lunar soil mixing models. *Proc. Second Lunar Sci. Conf., Suppl. 2, Geochim. Cosmochim. Acta, 2:* 1063-1081.

Gose, W. A., Pearce, G. W. and Lindsay, J. F., 1975. Magnetic stratigraphy of the Apollo 15 deep drill core. *Proc. Sixth Lunar Sci. Conf., Suppl. 6, Geochim. Cosmochim. Acta,* (in press).

Green, D. H., Ringwood, A. E., Ware, N. G., Hibberson, W. O., Major, H. and Kiss, E., 1971. Experimental petrology and petrogenesis of Apollo 12 basalts. *Proc. Second Lunar Sci. Conf., Suppl. 2, Geochim. Cosmochim. Acta, 1:* 601-615.

Halmos, P. R., 1944. Random alms. *Ann. Math. Stat., 15:* 182-189.

Heiken, G., Duke, M., Fryxell, R., Nagle, S., Scott, R. and Sellers, G., 1972. Stratigraphy of the Apollo 15 drill core. NASA TMX-58101, 21 pp.

Hodges, F. N. and Kushiro, I., 1973. Petrology of Apollo 16 lunar highland rocks. *Proc. Fourth Lunar Sci. Conf., Suppl. 4, Geochim. Cosmochim. Acta, 1:* 1033-1048.

Hubbard, N., Meyer, C., Gast, P. and Wiesmann, H., 1971. The composition and derivation of Apollo 12 soils. *Earth Planet. Sci. Lett., 10:* 341-350.

King, E. A., Butler, J. C. and Carman, M. F., 1971. The lunar regolith as sampled by Apollo 11 and 12; grain size analyses, and modal amalyses, origins of particles. *Proc. Second Lunar Sci. Conf., Suppl. 2, Geochim. Cosmochim. Acta, 1:* 737-746.

King, E. A., Butler, J. C. and Carman, M. F., 1972. Chondrules in Apollo 14 samples and size analyses of Apollo 14 and 15 fines. *Proc. Third Lunar Sci. Conf., Suppl. 3, Geochim. Cosmochim. Acta, 1:* 673-686.

Kolmogoroff, A. N., 1941. Uber das logarithmisch normale verteilungsgesetz der dimensionen der teilchen bel zerstuchelung. *Akad. Nauk SSSR, Doklady, 31:* 99-101.

Lindsay, J. F., 1971a. Sedimentology of Apollo 11 and 12 lunar soil. *Jour. Sediment. Pet., 41:* 780-797.

Lindsay, J. F., 1971b. Mixing models and the recognition of end-member groups in Apollo 11 and 12 soils. *Earth Planet. Sci. Lett., 12:* 67-72.

Lindsay, J. F., 1972a. Sedimentology of clastic rocks from the Fra Mauro region of the moon. *Jour. Sediment. Pet., 42:* 19-32.

Lindsay, J. F., 1972b. Sedimentology of clastic rocks returned from the moon by Apollo 15. *Geol. Soc. Amer. Bull., 83:* 2957-2970.

Lindsay, J. F., 1972c. Development of soil on the lunar surface. *Jour. Sediment. Pet., 42:* 876-888.

Lindsay, J. F., 1973. Evolution of lunar grain-size and shape parameters. *Proc. Fourth Lunar Sci. Conf., Suppl. 4, Geochim. Cosmochim. Acta, 1:* 215-224.

Lindsay, J. F., 1974. Transportation of detrital materials on the lunar surface: evidence from Apollo 15. *Sedimentology, 21:* 323-328.

Lindsay, J. F., 1975. A steady state model for the lunar soil. *Geol. Soc. Amer. Bull., 86:* 1661-1670.

Lindsay, J. F. and Srnka, L. J., 1975. Galactic dust lanes and the lunar soil. *Nature, 257:* 776-778.

Lindsay, J. F., Heiken, G. H. and Fryxell, R., 1971. Description of core samples returned by Apollo 12. NASA TMX-58066, 19 pp.

McKay, D. S. and Morrison, D. A., 1971. Lunar breccia. *Jour. Geophys. Res., 76:* 5658-5669.

McKay, D. S., Greenwood, W. R., and Morrison, D. A., 1970. Origin of small lunar particles and breccia from the Apollo 11 site. *Proc. Apollo 11 Lunar Sci. Conf., Suppl. 1, Geochim. Cosmochim. Acta, 1:* 673-693.

McKay, D. S., Morrison, D. A., Clanton, U. S., Ladle, G. H. and Lindsay, J. F., 1971. Apollo 12 soil and breccia. *Proc. Second Lunar Sci. Conf., Suppl. 2, Geochim. Cosmochim. Acta, 1:* 755-773.

McKay, D. S., Heiken, G. H., Taylor, R. M., Clanton, U. S. and Morrison, D. A., 1971. Apollo 14 soils. size distribution and particle types. *Proc. Third Lunar Sci. Conf., Suppl. 3, Geochim Cosmochim. Acta, 1:* 983-994.

McKay, D. S., Fruland, R. M. and Heiken, G. H., 1974. Grain size and the evolution of lunar soils. *Proc. Fifth Lunar Sci. Conf., Suppl. 5, Geochim. Cosmochim. Acta, 1:* 887-906.

Meyer, C., Brett, R., Hubbard, N., Morrison, D., McKay, D., Aitken, F., Takeda, H. and Schonfeld, E., 1971. The mineralogy chemistry and origin of the KREEP component in soil samples from the Ocean of Storms. *Proc. Second Lunar Sci. Conf., Suppl. 2, Geochim. Cosmochim. Acta, 1:* 393-411.

Moore, J. G. and Peck, D. L., 1962. Accretionary lapilli in volcanic rocks of the western continental United States. *Jour. Geol., 70:* 182-193.

Nakamura, Y., Dorman, J., Duennebier, F., Lammlein, D., and Latham, G., 1975. Shallow lunar structure determined from the passive seismic experiment. *The Moon, 13:* 57-66.

Oberbeck, V. R., and Quaide, W. L., 1967. Estimated thickness of a fragmental surface layer of Oceanus Procellarum. *Jour. Geophys. Res., 72:* 4697-4704.

Oberbeck, V. R. and Quaide, W. L., 1968. Genetic implications of lunar regolith thickness variations. *Icarus, 9:* 446-465.

Pettijohn, F. J., 1957. *Sedimentary Rocks.* Harper and Rowe, N. Y., 2nd Ed., 718 pp.

Quaide, W. L. and Oberbeck, V. R., 1968. Thickness determinations of the lunar surface layer from lunar impact craters. *Jour. Geophys. Res., 73:* 5247-5270.

Quaide, W. L. and Oberbeck, V. R., 1975. Development of mare regolith: some model implications. *The Moon, 13:* 27-55.

Reid, A. M., 1974. Rock types present in lunar highland soils. *The Moon, 9:* 141-146.

Reid, A. M., Warner, J., Ridley, W. I., Johnston, D. A., Harmon, R. S., Jakes, P. and Brown, R. W., 1972a. The major element compositions of lunar rocks as inferred from glass compositions in the lunar soils. *Proc. Third Lunar Sci. Conf., Suppl. 3, Geochim. Cosmochim. Acta, 1:* 379-399.

Reid, A. M., Warner, J., Ridley, W. I., and Brown, R. W., 1972b. Major element composition of glasses in three Apollo 15 soils. *Meteoritics, 7:* 395-415.

Rennilson, J. J., Dragg, J. L., Morris, E. C., Shoemaker, E. M. and Turkevich, A., 1966. Lunar surface topography. Surveyor I Mission Report, Part 2 Scientific Data and Results, *Jet Propulsion Lab. Tech. Rept. 82-1023.*

Rhodes, J. M., Rodgers, K. V., Shih, C., Bansal, B. M., Nyquist, L. E., Wiesmann, H., and Jubbard, N. J., 1974. The relationship between geology and soil chemistry at the Apollo 17 landing site. *Proc. Fifth Lunar Sci. Conf., Suppl. 5, Geochim. Cosmochim. Acta, 2:* 1097-1117.

Rhodes, J. M., Adams, J. B., Blanchard, D. P., Charette, M. P., Rodgers, K. V., Jocogs, J. W., Brannon, J. C. and Haskin, L. A., 1975. Chemistry of agglutinate fractions in lunar soils. *Proc. Sixth Lunar Sci. Conf., Suppl. 6, Geochim. Cosmochim. Acta,* (in press).

Ridley, W. I., Reid, A. M., Warner, J. L., Brown, R. W., Gooley, R. and Donaldson, C., 1973. Glass compositions in Apollo 16 soils 60501 and 61221. *Proc. Fourth Lunar Sci. Conf., Suppl. 4, Geochim. Cosmochim. Acta, 1:* 309-321.

Ringwood, A. E. and Essene, E., 1970. Petrogenesis of Apollo 11 basalts, internal constitution and origin of the moon. *Proc. Apollo 11 Lunar Sci. Conf., Suppl. 1, Geochim. Cosmochim. Acta, 1*: 769-799.

Schonfeld, E. and Meyer, C., 1972. The abundances of components of the lunar soils by a least-squares mixing model and the formation age of KREEP. *Proc. Third Lunar Sci. Conf., Suppl. 3, Geochim. Cosmochim. Acta, 2*: 1397-1420.

Scott, R. F. and Robertson, F. I., 1967. Soil mechanics surface samples; Lunar surface tests, results and analyses: Surveyor III Mission Report, Part II, Scientific Results. *Jet Propulsion Lab. Tech. Rept. 82-1177*, 69-110.

Shoemaker, E. M., Batson, R. M., Holt, H. E., Morris, E. C., Rennilson, J. J. and Whitaker, E. A., 1967. Television observations for Surveyor III: Surveyor III Mission Report, Part II, Scientific Results. *Jet Propulsion Lab. Tech. Rept. 82-1177*, 9-67.

Shoemaker, E. M., Hait, M. H., Swann, G. A., Schleicter, D. L., Schaber, G. G., Sutton, R. L., Dahlem, D. H., Goddard, E. N. and Waters, A. C., 1970. Origin of the lunar regolith at Tranquility Base. *Proc. Apollo 11 Lunar Sci. Conf., Suppl. 1, Geochim. Cosmochim. Acta, 3*: 2399-2412.

Simonds, C. H., 1973. Sintering and hot pressing of Fra Mauro composition glass and the lithification of lunar breccias. *Am. Jour. Sci., 273*: 428-439.

Taylor, S. R., 1975. *Lunar Science: A Post-Apollo View*. Pergamon Press, N. Y., 372 pp.

Turkevich, A. L., 1973a. The average chemical composition of the lunar surface. *Proc. Fourth Lunar Sci. Conf., Suppl. 4, Geochim. Cosmochim. Acta, 2*: 1159-1168.

Turkevich, A. L., 1973b. Average chemical composition of the lunar surface. *The Moon, 8*: 365-367.

Uhlmann, D. R., Klein, L. and Hopper, R. W., 1975. Sintering, crystallization, and breccia formation. *The Moon, 13*: 277-284.

Walker, D., Grove, T. L., Longhi, J., Stolper, E. M. and Hays, J. F., 1973. Origin of lunar feldspathic rocks. *Earth Planet. Sci. Lett., 20*: 325-336.

Waters, A. C., Fisher, R. V., Garrison, R. E. and Wax, D., 1971. Matrix characteristics and origin of lunar breccia samples no. 12034 and 12037. *Proc. Second Lunar Sci. Conf., Suppl. 2, Geochim. Cosmochim. Acta, 1*: 893-907.

Watkins, J. S. and Kovach, R. L., 1973. Seismic investigations of the lunar regolith. *Proc. Fourth Lunar Sci. Conf., Suppl. 4, Geochim. Cosmochim. Acta, 3*: 2561-2574.

Wood, J. A., Dickey, J. S., Marvin, U. B. and Powell, B. N., 1970. Lunar anorthosites and a geophysical model of the moon. *Proc. Apollo 11 Lunar Sci. Conf., Suppl. 1, Geochim. Cosmochim. Acta, 1*: 965-988.

Lunar Glossary

Note to the User:

A number of terms are unique to lunar science or are at least used in a specialized sense. The following brief glossary is an attempt to define these unique terms plus provide definitions of terms which are only rarely used in the geologic literature. In part the glossary is adapted from the Proceedings of the Fifth Lunar Science Conference (Vol. 1) and has been modified to suit the needs of this text.

Glossary

Accretion. The process by which planetary bodies increase in size by incorporating interplanetary material.

Aeon. 10^9 yr.

Agglutinate. A common particle type in lunar soils, agglutinates consist of comminuted rock, mineral and glass fragments bonded together with glass. The glass is black or dark brown in bulk, but pale brown to very dark brown in thin section, and is characteristically heterogeneous, with dark brown to black flow banding or "schlieren".

Albedo. The ratio of the brightness of a reflecting object to that of a theoretical perfectly diffusing (i.e., obeying Lambert's Law) flat surface at the same position and having the same projected surface area.

Alkalic high-alumina basalt. Lunar rocks with 45—60% modal or normative plagioclase; mafic minerals predominantly low Ca pyroxene with varying amounts of olivine; total alkalis and phosphorus are relatively high. Compositionally equivalent to KREEP basalt.

Anorthosite. Term used for lunar rock with over 90% modal or normative plagioclase. Has also been used rather loosely to encompass all feldspathic rocks in the lunar highlands.

Anorthositic gabbro. Used for lunar rocks with 65—77.5% modal per cent plagioclase (normative plagioclase has been used to classify particles from compositional data).

Anorthositic norite. Used for lunar rocks with 60–77.5% modal or normative plagioclase; low Ca pyroxene dominant over high Ca pyroxene.

Armalcolite. New lunar mineral ($Mg_{x>1} Fe_{1-x} Ti_2O_5$) that is isomorphous with ferropseudobrookite and karooite. It is generally interpreted as a primary phase and is commonly rimmed by ilmenite.

Basalt. Fine grained, commonly extrusive, mafic igneous rock composed chiefly of calcic plagioclase and clinopyroxene in a glassy or fine-grained groundmass. Lunar basalts contain plagioclase of bytownitic or anorthitic composition and ilmenite as a major phase. Term is also used in a purely compositional sense for lithic fragments and glasses.

Base surge. A debris cloud near the ground surface that moves radially from a chemical, nuclear or volcanic explosive center or from a meteoritic impact center. The first use was to describe clouds at the base of shallow thermonuclear explosions. Lunar applications of the term are based on photogeologic interpretations of patterned ground or dunes around large lunar craters such as Tycho. Accretionary structures in soil breccias have been interpreted as having formed in impact-generated, hot base surge clouds.

Breccia. Clastic rock composed of angular clasts cemented together in a finer-grained matrix.

Cataclastic. A metamorphic texture produced by mechanical crushing, characterized by granular, fragmentary, or strained crystals.

Cumulate. A plutonic igneous rock composed chiefly of crystals accumulated by sinking or floating from a magma. Terrestrial cumulates are usually structuraly layered (layers varying in the proportion of cumulus phases) but textural criteria could also be applied.

Dark matrix breccia. Polymict breccia with dark-colored glassy or fine-grained matrix. Used specifically for breccias containing lithic clasts angular to spherical glass fragments, and single crystals in a matrix of brown glass.

Diaplectic glass. Glass formed in the solid state from a single mineral grain due to the passage of a shock wave.

Dunite. Used for lunar rocks with over 90% modal or normative olivine.

Ejecta. Materials ejected from the immediate crater by a volcanic explosion or meteoroid impact.

Exposure age. Period of time during which a sample has been at or near the lunar surface, assessed on the basis of cosmogenic rare gas contents, particle track densities, short-lived radioisotopes, or agglutinate contents in the case of soil samples.

Fra Mauro basalt. Defined originally from glass compositions in soils at the

Apollo 14 site. A basaltic composition with the approximate composition (in weight per cent) SiO_2 48, TiO_2 2, Al_2O_3 18, FeO 10, MgO 9, CaO 11, Na_2O 0.7, K_2O 0.5, P_2O_5 1. Compositionally equivalent to KREEP basalt.

Grabbroic anorthosite. Used for lunar rocks with 77.5–90% modal or normative plagioclase (normative plagioclase from compositional data has been used to classify lithic fragments and glasses).

Granite. Compositional term used for lunar rocks or glasses rich in Si and K. Common composition is approximately SiO_2 70, Al_2O_5 14, FeO 2, MgO 1, CaO 2, Na_2O 1, K_2O 7 (weight per cent). The term *rhyolite* has been used in the same compositional sense.

High-alumina basalt. Lunar rocks with 45–60% modal (or normative) plagioclase; mafic minerals predominantly low Ca pyroxene with varying amounts of olivine. Used to encompass both low-alkali high-alumina basalt and alkalic high-alumina basalt. Commonly also used for high-alumina basalt compositions like KREEP basalt but with substantially lower K and P; i.e. equivalent to low-alkali, high-alumina basalt. Not equivalent to terrestrial high-alumina basalt.

Highland basalt. Compositional term for rocks or glasses with the composition of very aluminous basalt (approximate composition SiO_2 45, TiO_2 .4, Al_2O_3 26, FeO 6, MgO 8, CaO 15, Na_2O .3, K_2O <.1 weight per cent).

Hornfelsic. A fine-grained metamorphic texture of equidimensional grains with no preferred orientation. Porphyroblasts or relict phenocrysts may be present.

Impact melt. Melt produced by fusion of target rock due to impact of a meteoroid.

Intersertal. A groundmass texture in a porphyritic rock in which unoriented feldspar laths enclose glassy or partly crystalline material other than augite.

Light matrix material. Breccias with light colored feldspathic matrices. Used specifically for feldspathic breccias with anorthositic bulk compositions and generally with some recrystallization, or with unrecrystallized matrices.

Mare basalt. Basaltic igneous rocks from the mare regions of the moon characterized by high FeO (>14), commonly high TiO_2, low Al_2O_3 (<11) and low-alkali contents. Major minerals are clinopyroxene and calcic plagioclase, with lesser Fe-Ti-Cr oxides, metallic iron, and troilite. Olivine or a SiO_2 polymorph or both are common.

Mascon. Concentration of mass, apparently at a relatively shallow depth within the moon, commonly associated with circular mare basins.

Maskelynite. Plagioclase that has been transformed by shock in the solid state to a glass. Equivalent to diaplectic or thetomorphic plagioclase glass.

Meteoritic component. These portions of lunar samples contributed by meteoroids which impacted after early lunar differentiation patterns had become established; can be detected through siderophile and volatile element abundance patterns.

Monomict breccia. A breccia formed by fracturing and mixing of material from a single source without admixture of unrelated material (cf. *Polymict breccia*).

Norite. Rock of basic composition consisting essentially of plagioclase and orthopyroxene; clinopyroxene should not exceed half of the total pyroxene content. Term has been used for a variety of lunar rocks that are generally basaltic in composition with orthopyroxene as a major phase; also used for basaltic compositions in which the normative pyroxene is low in Ca.

Noritic anorthosite. Used for lunar rocks with 77.5—90% modal or normative plagioclase; low Ca pyroxene dominant over high Ca pyroxene.

Oikocrysts. In a poikilitic or poikiloblastic texture, the crystal that encloses the chadacrysts.

Ophitic. A basaltic texture characterized by laths of plagioclase partially enclosed by anhedral grains of pyroxene. Believed to represent comtemporaneous crystallization of the two minerals, rather than a sequence as in poikilitic texture.

Polymict breccia. A breccia containing fragments of different compositions and origins (cf. *Monomict breccia*).

Regolith. Lunar regolith is the fragmental debris, produced principally by impact processes, which lies on bedrock.

Regolith breccia. See *Soil breccia.*

Rhyolite. See *Granite.*

Rille. Valleys on the lunar surface; classified by shape as straight, arcuate, or sinuous.

Secondary crater. Crater produced by impact of a projectile ejected from the lunar surface upon meteoroid impact.

Soil. The term soil may . . . "be loosely applied to all unconsolidated material above bedrock that has been in any altered or weathered". Lunar soils are primarily derived by the physical process of impact comminution and impact fusion of detrital particles.

Soil breccia. Polymict breccia composed of cemented or sintered lunar soil.

Soil maturity. Maturity of a soil refers to the degree of reworking by micrometeoroids, as evidenced by glass content, grain size parameters, relative amount of grains with high solar flare track densities, solar wind gas content, and other parameters shown to be related to the time of exposure at the lunar surface.

Troctolite. Terms for lunar rocks consisting essentially of plagioclase and olivine with little or no pyroxene. If spinel-bearing, it is termed spinel-troctolite.

Vitrification. Formation of a glass from a crystalline precursor generally by impact melting.

Index